Switching to ArcGIS® Pro
from ArcMap™

MARIBETH H. PRICE

Esri Press
REDLANDS | CALIFORNIA

Esri Press, 380 New York Street, Redlands, California 92373-8100
Copyright 2019 Esri
All rights reserved
22 21 20 19   2 3 4 5 6 7 8 9 10
Printed in the United States of America

Library of Congress Cataloging-in-Publication Data
Names: Price, Maribeth Hughett, 1963- author.
Title: Switching to ArcGIS Pro from ArcMap / Maribeth H. Price.
Description: Redlands, California : Esri Press, [2019] | Includes
    bibliographical references and index.
Identifiers: LCCN 2018046716 (print) | LCCN 2018052050 (ebook) | ISBN
    9781589485457 (electronic) | ISBN 9781589485440 (pbk. : alk. paper)
Subjects: LCSH: ArcGIS. | ArcMap. | Geographic information systems. |
    Graphical user interfaces (Computer systems)
Classification: LCC G70.212 (ebook) | LCC G70.212 .P745 2019 (print) | DDC
    910.285--dc23
LC record available at https://urldefense.proofpoint.com/v2/url?u=https-
3A__lccn.loc.gov_2018046716&d=DwIFAg&c=n6-cguzQvX_tUIrZOS_4Og&r=qNU49__SCQN30XC-f38qj8bYYMTIH4VCOt2
Jb8fvjUA&m=avUyJlwaNq6hHxeR92GbpcvN2nmPFUEALrVxgXkzWIc&s=nBly3EirNZZEqy7hLAYSdDGqPOajx4fRC8bRfaU2T
J0&e=

In North America:
Ingram Publisher Services
Toll-free telephone: 800-648-3104
Toll-free fax: 800-838-1149
E-mail: customerservice@ingrampublisherservices.com

In the United Kingdom, Europe, the Middle East and Africa, Asia, and Australia:
Eurospan Group   Telephone 44(0) 1767 604972
3 Henrietta Street   Fax: 44(0) 1767 6016-40
London WC2E 8LU   E-mail:eurospan@turpin-distribution.com
United Kingdom

Contents

Preface .... *ix*

1 Contemplating the switch to ArcGIS Pro **1**
  Background 1
    System requirements 2
    Licensing 3
    Capabilities of ArcGIS Pro 4
    When should I switch? 5
  Time to explore **8**
    Objective 1.1: Downloading the data for these exercises 8
    Objective 1.2: Starting ArcGIS Pro, signing in, creating a project, and exploring the interface .. 10
    Objective 1.3: Accessing maps and data from ArcGIS Online 14
    Objective 1.4: Arranging the windows and panes 17
    Objective 1.5: Accessing the help 18
    Objective 1.6: Importing a map document 19

2 Unpacking the GUI **21**
  Background **21**
    The ribbon and tabs 21
    Panes 22
    Views 23
  Time to explore **24**
    Objective 2.1: Getting familiar with the Contents pane 24
    Objective 2.2: Learning to work with objects and tabs 27
    Objective 2.3: Exploring the Catalog pane 29

3 The project **33**
  Background **33**
    What is a project? 33
    Items stored in a project 34
    Paths in projects 36
    Renaming projects 37

Time to explore 38

    Objective 3.1: Exploring different elements of a project 38

    Objective 3.2: Accessing properties of projects, maps, and other items 41

4   Navigating and exploring maps **43**

  Background **43**

    Exploring maps 43

    2D and 3D navigation 44

  Time to explore **46**

    Objective 4.1: Learning to use the Map tools 46

    Objective 4.2: Exploring 3D scenes and linking views 50

5   Symbolizing maps **53**

  Background **53**

    Accessing the symbol settings for layers... 53

    Accessing the labeling properties.. 54

    Symbolizing rasters 56

  Time to explore **57**

    Objective 5.1: Modifying single symbols 57

    Objective 5.2: Creating maps from attributes.. 60

    Objective 5.3: Creating labels 63

    Objective 5.4: Managing labels 66

    Objective 5.5: Symbolizing rasters 68

6   Geoprocessing **71**

  Background **71**

    What's different.. 71

    Analysis buttons and tools.. 73

    Tool licensing 73

  Time to explore **74**

    Objective 6.1: Getting familiar with the geoprocessing interface 74

    Objective 6.2: Performing interactive selections 76

    Objective 6.3: Performing selections based on attributes 78

    Objective 6.4: Performing selections based on location 80

    Objective 6.5: Practicing geoprocessing 81

7   Tables **83**

  Background **83**

    General table characteristics... 83

    Joining and relating tables 84

    Making charts... 87

  Time to explore **88**

    Objective 7.1: Managing table views.. 88

Objective 7.2: Creating and managing properties of a chart 90
Objective 7.3: Calculating statistics for tables 92
Objective 7.4: Calculating and editing in tables.. 95

8   Layouts 97
Background... 97
Layouts and map frames..97
Layout editing procedures 98
Importing map documents and templates 100
Time to explore 101
Objective 8.1: Creating the maps for the layout.. 101
Objective 8.2: Setting up a layout page with map frames... 102
Objective 8.3: Setting map frame extent and scale 103
Objective 8.4: Formatting the map frame 104
Objective 8.5: Creating and formatting map elements... 106
Objective 8.6: Fine-tuning the legend 108
Objective 8.7: Accessing and copying layouts 110

9   Managing data 111
Background.. 111
Data models 111
Managing the geodatabase schema 112
Creating domains.. 115
Managing data from diverse sources 116
Project longevity 116
Managing shared data for work groups.. 117
Time to explore 118
Objective 9.1: Creating a project and exporting data to it 118
Objective 9.2: Creating feature classes... 120
Objective 9.3: Creating and managing metadata... 122
Objective 9.4: Creating fields and domains 124
Objective 9.5: Modifying the table schema 126
Objective 9.6: Sharing data using ArcGIS Online... 128

10 Editing 131
Background.. 131
Basic editing functions.. 131
Creating features.. 133
Modifying existing features 134
Creating and editing annotation 136
Time to explore 136
Objective 10.1: Understanding the editing tools in ArcGIS Pro.................................... 136
Objective 10.2: Creating points 138

Objective 10.3: Creating lines 140

Objective 10.4: Creating polygons 142

Objective 10.5: Modifying existing features 144

Objective 10.6: Creating an annotation feature class 146

Objective 10.7: Editing annotation 148

Objective 10.8: Creating annotation features 149

11 Moving forward 153

Data sources 155

Index 157

Preface

Change is never easy, especially when an accustomed way of doing things is challenged by new methods. This book was written to help ArcMap™ users adjust to the new ArcGIS® Pro software. Ten of its 11 chapters cover the most fundamental and commonly used aspects of geographic information systems (GIS). Each chapter includes an introductory discussion of the salient changes, followed by a set of practical hands-on exercises to lead the reader through the process of learning ArcGIS Pro.

ArcGIS®, ArcMap™, ArcCatalog™, ArcInfo®, ArcGIS® Online, and ArcGIS® Pro are trademarks of Esri®. The trademark symbols have been omitted hereafter for ease of reading, but no infringement of rights is hereby intended or condoned.

As the title indicates, this book is written for GIS professionals who already know ArcMap and have significant experience using it. It is not designed to teach beginning users of GIS as it assumes prior knowledge of the terminology, data structures, and procedures encountered by users of ArcGIS software. Rather than teaching each topic from the basics, it focuses on how ArcGIS Pro is different from ArcMap. It aims to quickly enable someone to make the transition, and it can be completed in a few days of serious effort, although additional practice will be needed for the new program to become second nature.

Even experienced users may find their knowledge of certain topics on the weak side. Consulting either ArcGIS Desktop Help for ArcMap or ArcGIS Pro Help for ArcGIS Pro will, in most cases, provide enough background to continue with the exercises.

This book assumes that the user has a license for ArcGIS Pro. An organizational account for ArcGIS Online is recommended, and usually an account will be integrated with the user's license for ArcGIS Pro. If not, an ArcGIS℠ Online Public Account may be used for the few exercises that require an account. It does not require access to any additional extensions such as ArcGIS 3D Analyst™ or ArcGIS Spatial Analyst.

The data used in the lessons is freely available and downloadable online and requires approximately 65 MB of space. All the data has been compiled from public sources and is redistributable with attribution to its creators. For information on how to download the data, go to objective 1.1 in chapter 1.

Acknowledgments

The author thanks Esri for permission to use the screen shots of the ArcGIS Pro GUI, ArcGIS Pro Help and ArcGIS Desktop Help, and basemaps reproduced in the text. Used with permission. Copyright © 2018 Esri, Crater Lake National Park, National Park Service, ArcGIS, ArcGIS Pro, ArcMap, *Geologic Map of the Edwards Aquifer Recharge Zone*, US Geological Survey (USGS), INCREMENT P, © OpenStreetMap, US Geological Survey Scientific Investigations, EarthExplorer, contributors, and the GIS user community. All rights reserved.

Early users of this book and workshop attendees have contributed to the book's improvement.

The author also thanks the South Dakota School of Mines and Technology for a career in teaching GIS, and the many students who challenged me to continually find better ways to explain GIS concepts. Their dedication and enthusiasm have been a lifelong inspiration.

Chapter 1

Contemplating the switch to ArcGIS Pro

Background

ArcGIS Pro will seem both familiar and completely new to users of ArcMap and ArcCatalog. The GUI arrangement echoes the design of ArcMap: the map in the middle, map layers on one side, the data catalog on the other, with menus above (figure 1.1). Much of the architecture and terminology will persist in the new paradigm: geographic data, maps, layouts, geoprocessing tools, tables, joins, and so on. However, the implementation and details of how these functions are accessed and manipulated takes a completely different approach in some cases. For users who know ArcMap well, the experience will be initially frustrating as one searches for familiar tools and tasks that no longer appear in the same spot, or as one gets used to a totally new mindset for some tasks. Too, ArcGIS Pro has only gradually developed the capabilities already available in ArcMap, so it is possible that the feature sought has not yet been implemented.

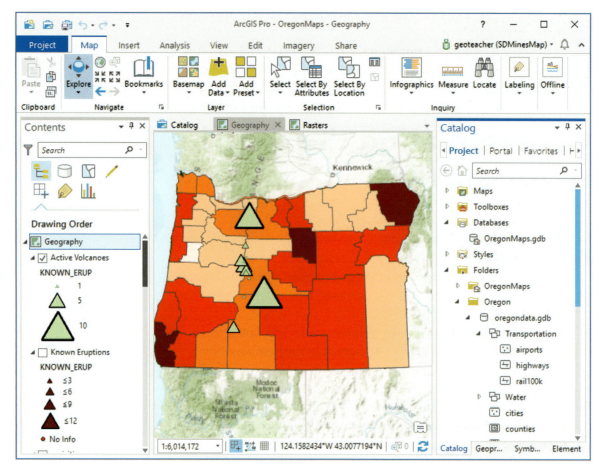

Figure 1.1. ArcGIS Pro showing a map of Oregon.

However, tapping down the chagrin and opening up to a new way of doing certain things will have its rewards. New users unfamiliar with ArcMap seem to learn ArcGIS Pro more quickly as it follows modern software GUI conventions (a ribbon with tabs and context-sensitive menus). In many aspects, it is more intuitive and more efficient. After some months of using ArcGIS Pro exclusively, I had to return to ArcMap for a month, and my immediate reaction was "Oh my, how clunky." Hopefully after some practice, you too will grow to like the new software.

System requirements

ArcGIS Pro requires a 64-bit multiprocessor machine with at least a dual core, and a quad-core processor or higher is recommended. Unlike ArcMap, ArcGIS Pro operates in a multithreaded manner, enabling it to take full advantage of multicore computers, which ArcMap could not. Significant performance improvements should be expected in most cases. ArcGIS Pro operates better when more cores are available to it.

ArcGIS Pro requires a minimum of 8 GB of RAM and a good graphics card with at least 2 GB of RAM and 24-bit display color depth. For a complete description of system requirements, consult the Esri product website, http://pro.arcgis.com/en/pro-app/get-started/arcgis-pro-system-requirements.htm.

ArcGIS Pro is tightly joined with ArcGIS Online, a cloud-based system built to encourage sharing of GIS data, workflows, and other resources between organizations and users. It is strongly recommended that the user have an ArcGIS Online organizational account, which provides access to significantly more data and tools. ArcGIS Pro is also designed to run with a fast internet connection. Although the software can be run offline, provided the data sources are stored on the local computer, it will temporarily lose access to basemaps and internet-provided map services. If you're planning to use ArcGIS Pro offline regularly, such as on a field laptop, it makes sense to limit maps and tools to use only local data on the computer's hard drive.

Licensing

Because ArcGIS Pro is designed to interact with ArcGIS Online, its preferred method of licensing is an ArcGIS Online organizational account. In this method, the administrator of an ArcGIS Online organization creates a user account and assigns to it an ArcGIS Pro license from among the total number allotted to the organization. Extensions may also be assigned if available. Figure 1.2 shows the screen the ArcGIS Online administrator uses to assign licenses and extensions to a user account. When starting ArcGIS Pro, users are asked to sign into ArcGIS Online using their account, which simultaneously licenses the software and provides users access to any content they have created within ArcGIS Online.

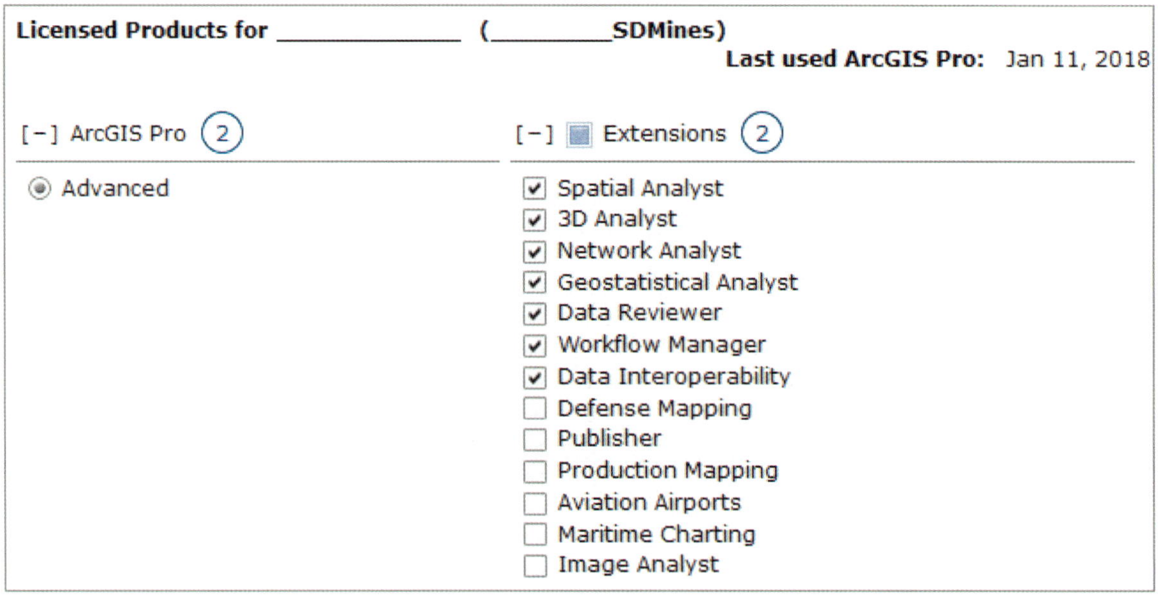

Figure 1.2. Configuring an ArcGIS Pro license for a user's ArcGIS Online account.

For continuity with previous licensing methods, Esri also offers a concurrent licensing model and a single-use license model, both previously used for ArcMap licensing. In the concurrent model, a license manager program runs on a server and allocates licenses at a user's request. This model is often called a "floating license," because it is assigned, and then taken back when the user is finished, allowing a fixed number of licenses to be shared across many computers. In the single-use model, the license is tied to a specific computer.

Capabilities of ArcGIS Pro

Although it is different in many ways, ArcGIS Pro will eventually have the equivalent capabilities of ArcMap and more. This updating process is ongoing, with new capabilities being added with each release. ArcGIS Pro 2.3, for which this book is tested, represented a significant milestone in which nearly all of ArcMap's key functions were present, and additional capabilities and improvements are still being added.

ArcGIS Pro integrates the functions of ArcMap, ArcCatalog, ArcToolbox, ArcScene, and ArcGlobe within a single interface so that 3D data visualization and analysis no longer require opening a separate program. ArcGIS Pro retains the Python® and ModelBuilder™ customization modules. Certain aspects that users are accustomed to using as separate features, such as COGO or Maplex™ for ArcGIS®, are now integrated directly into the program. The familiar extensions, such as Spatial Analyst or 3D Analyst, continue to be available and work in much the same way, as additional licensed tools in ArcToolbox.

However, users should be aware that the very design of ArcGIS Pro precludes certain familiar aspects of ArcMap from working precisely the same way. In ArcGIS Pro, the design criteria specify that most actions are geoprocessing events (taking advantage of the multiprocessor computer). Tools have replaced some of the common right-click commands, either as a temporary measure until the functionality is integrated into the GUI, or as a design choice. For example, instead of opening a window to execute a selection using table attributes with the Query Builder, you must run a tool named Select Layer By Attribute.

TIP *Can't find something? Try searching for a tool.*

When should I switch?

Several factors will affect the decision of when to switch to ArcGIS Pro.

Reasons for making the switch include:
- Improved performance that takes advantage of multicore architecture
- Seamless integration of two- and three-dimensional visualization, analysis, and editing
- Improved flexibility in creating dynamic labels
- Creating multiple layouts based on the same map or maps
- Tight integration with ArcGIS Online, making it easier to share maps and map services
- A more modern and intuitive GUI
- Expanded use of defaults that can make life easier for inexperienced users
- The product life cycle of ArcMap (See the Esri Support Product Life Cycles website.)

ArcMap users may dispute the "more intuitive interface" claim when encountering ArcGIS Pro for the first time. In truth, the GUI is probably more difficult to learn if you already know ArcMap and have become accustomed to certain ways of doing things. In general, new users find the ArcGIS Pro GUI easier to learn. Over time, you get used to it.

The software also implements reasonable defaults more than ArcMap, hiding complications from beginners. This feature has both benefits and pitfalls. For example, ArcGIS Pro does not automatically warn users when two different geographic coordinate systems (datums) are being used in the same map (although this warning can be turned on). This choice makes little difference unless the user is working at very large scales with precise data, and in many situations the issue can be safely ignored. However, it can cause difficulties later if inexperienced users compile data for a project without being aware that the feature classes and rasters use multiple datums.

When ArcMap retires, the software will continue to be available, and people can continue to use it, but it will no longer be updated or supported. A similar process occurred when ArcInfo® Workstation and ArcView became ArcGIS® Desktop.

Delaying the switch might be wise under some circumstances:
- The organization does not have enough computers capable of running ArcGIS Pro.
- Some critical functionality may not yet be available.
- The organization has many custom scripts or tools that must be updated and tested.
- Third-party extensions (such as HEC-RAS or MODFLOW) may not yet be supported.
- Maps and layouts must be shared with ArcMap users.

Several other considerations regarding data must also be considered:
- ArcGIS Pro does not recognize personal geodatabases or ArcInfo coverages.
- The organization and structure of maps and layouts are different.
- Data management in ArcGIS Pro has complications not seen in ArcCatalog.
- Absolute/relative paths are now fixed by default and have less flexibility.
- The Excel™ interpreter is different and seems less forgiving than the one in ArcMap.

As discussed in the next section, even though the data format is consistent (the geodatabase), the way that files are organized is quite altered and not as flexible. To switch to ArcGIS Pro, you must learn different ways of doing things. For example, it is extremely difficult to rename a project, so you are stuck with the initial name, even if you decide it is inappropriate or confusing. You no longer have a choice whether to use absolute or relative paths in a document; the choice is made by the software. This feature removes a potential trap for inexperienced users, and the defaults usually work well, but experienced users who are accustomed to controlling this option may find that they need to change how they organize datasets.

If you use many Excel files, problems may occur. The Excel interpreter has had difficulty reading files, in my experience, and usually I must resort to the comma-delimited CSV file format instead. The errors can be delayed and subtle. For example, in one project an Excel table imported just fine but generated fatal errors when it was exported to a geodatabase. A CSV version of the same table was imported and exported to the geodatabase without trouble.

Mapping differences to consider before switching:
- The default symbols follow a different cartographic style.
- Previous custom styles (such as Civic or Forestry) are not included and must be imported.
- Labels are more flexible, but the options are more numerous and take getting used to.
- Annotation editing tools are still catching up to ArcMap in functionality and reliability.

The symbol styles have been changed to favor web mapping applications and may make it difficult to see multiple layers well. Dark colors with white outlines dominate the default symbols, and although they can be changed, it adds an extra step when a feature class is added to a map. Previous custom style sets must be imported from the ArcMap symbol folder, although if a map document is imported, the symbols will be included. In general, the map document import function works well, although a few settings may not be preserved and will need to be reset, such as the data frame clip setting.

Annotation can no longer be saved in a map as graphic text; it can only be created as a geodatabase feature class. The annotation editing tools have lagged in development. ArcGIS Pro 2.0 had one annotation construction tool (Straight), and ArcGIS Pro 2.3 has three (Horizontal, Straight, and Curved). The workflow for placing unplaced annotation appears to be a manual process of editing the attribute table, without the convenient overflow window. Users who rely heavily on annotation may wish to stick to ArcMap a while longer.

Geoprocessing considerations to consider before switching:
- Some familiar menu tasks have been replaced by tools.
- Tool outputs are now saved to a project geodatabase by default.
- Background processing is "gone."

Overall, geoprocessing is similar, except that familiar functions, such as joining two tables, may now require running a tool instead of opening a menu. The spatial join, available in ArcMap as a relatively simple window, must now be run as the full tool, and although the tool has more options, it is complex compared with the simple menu in ArcMap.

The background geoprocessing option has also disappeared as a separate option, but only in a sense. ArcGIS Pro, with its multithreaded design, essentially runs all tools in the background, and it does so more quickly and reliably than the ArcMap background geoprocessor.

All these factors and considerations will influence an organization in deciding when it is time to switch to the new software. Because ArcMap/ArcCatalog and ArcGIS Pro can be installed and run on the same computer, users have more time and flexibility when making the switch. (However, it is not recommended to run both programs at the same time: both are memory hogs, and issues may result if they try to modify the same data simultaneously). They use the same data models so that you can work with the same datasets using either program. However, maps and layouts are completely different. You can import an ArcMap map document to ArcGIS Pro and save it to the new format, but ArcGIS Pro cannot edit map documents, and ArcMap cannot work with ArcGIS Pro layouts. Once created, an ArcGIS Pro layout cannot be saved to the older map document format.

Time to explore

This exercise assumes that you have installed ArcGIS Pro, have been assigned an ArcGIS Pro license, and possess an ArcGIS Online organizational account. It is helpful to be familiar with the ArcGIS Online terminology for services, such as web maps, feature layers, and so on.

Objective 1.1: Downloading the data for these exercises

A set of data has been assembled to use with this book, and the data must be downloaded and installed on your computer. It is smart to keep data organized. It is best to put GIS data on drive C; the Desktop and user Documents folders don't always work well, especially in networked environments. You can download the data from ArcGIS Online, as described next.

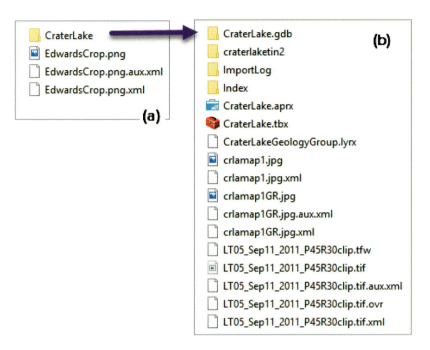

Figure 1.3. Contents of the SwitchToProData folder as shown in Windows Explorer.

The data is saved to a book group named Switch to ArcGIS Pro (Esri Press) in the Learn ArcGIS organization.

1. **Go to https://www.arcgis.com and log in with an ArcGIS Online account.**

2. **On the Home tab, in the Search box, type** Switch to ArcGIS Pro, **and then click the Search for Groups entry in the list. If no groups are found, turn off the option to "Only search in (your organization)."**

3. **Click the link to open the Switch to ArcGIS Pro (Esri Press) group and find the data, a zip file named SwitchToProData.**

4. **Click the thumbnail and download the file, saving it to a location on drive C, rather than in the Documents library or on Desktop.**

5. **Extract the zip file. It will create a folder named SwitchToProData.** The SwitchToProData folder contains an ArcGIS Pro project named CraterLake and a PNG image named EdwardsCrop (figure 1.3a). Within the CraterLake project folder (figure 1.3b) is a file geodatabase, a TIN, a project file (.aprx), a layer file (.lyrx), a toolbox, a couple of JPEG (.jpg) images, and a multiband Landsat 5 scene.

Objective 1.2: Starting ArcGIS Pro, signing in, creating a project, and exploring the interface

The preferred licensing model for ArcGIS Pro uses a named user account, which is an ArcGIS Online subscription organizational account that your administrator has configured with an ArcGIS Pro license.

Instead of a map document used by ArcMap, ArcGIS Pro uses a more complex data structure called a *project*, which will be presented in detail in chapter 3.

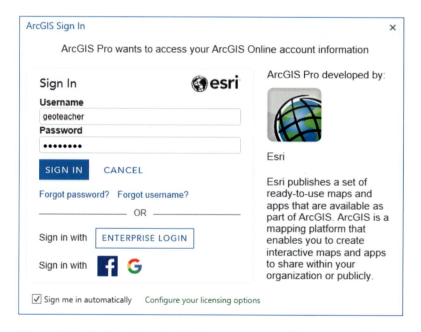

Figure 1.4. The first time ArcGIS Pro starts, you will be prompted to enter your ArcGIS Online user name and password.

Figure 1.5. The initial start screen of ArcGIS Pro.

1. **Start ArcGIS Pro (figure 1.4) and log in using your ArcGIS Online account.** The initial start screen for ArcGIS Pro presents a variety of options (figure 1.5). On the left, you can open a project that is already saved by choosing one from the recent list or a pinned list, or by browsing for another one. The middle section is used to create a new project using one of the system templates: Map, Catalog, Global Scene, or Local Scene. The key difference between them is the initial type of display that will be created. The section on the right lets you access your own templates.

2. **In the middle section, under the heading Create a project from a system template, choose the Catalog template (figure 1.5).**

3. **Enter** LearnPro **for the project name (figure 1.6). Click the Browse button to specify the new SwitchToProData folder as the location. Keep the option to create a new folder, and finish creating the project.**

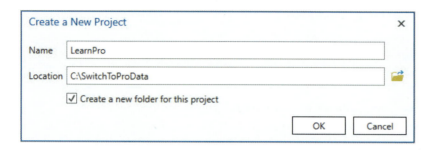

Figure 1.6. Creating a new, blank project named LearnPro.

4. **Examine the interface.**

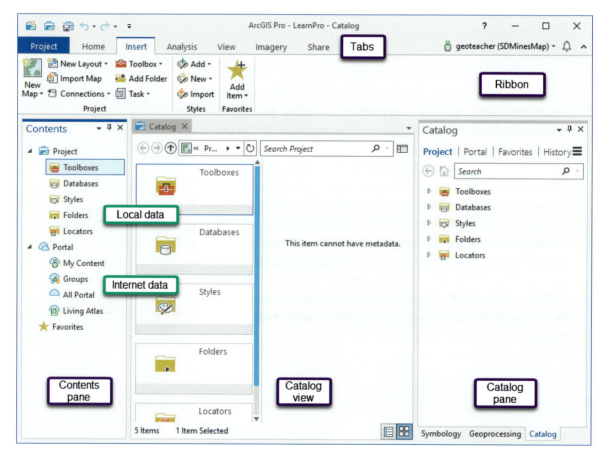

Figure 1.7. The ArcGIS Pro GUI showing a new project.

TIP If your interface looks different from figure 1.7, close every window using the X in the upper-right corner, until only the tabs remain. Then open the View tab and click the Contents button, the Catalog Pane button, and the Catalog View button (preferably in that order). You can use the View tab at any time if you inadvertently close one of these main panes.

5. **Examine the top of the GUI containing the tabs used to access features and functions.**

6. **Review the Contents pane (probably on the left side).** This pane behaves much like the ArcMap Table of Contents.

    Unlike the Table of Contents in ArcMap, which shows only map layers, the Contents pane in ArcGIS Pro portrays other types of content as well, as it does here in showing the contents of the project.

    The middle section shows the Catalog view, which portrays the metadata for, or a preview of, items selected in the Contents pane. It is also used to edit metadata.

7. **Click several entries in the Contents pane. The Catalog view updates to show the items contained within the clicked item.** Note that in the top half of the Contents pane, the Project folder links to resources saved to the computer. The bottom half, the Portal folder, accesses internet servers.

8. **Peruse the Catalog pane on the right side; it is analogous to the Catalog window in ArcMap. Although it is similar to the Catalog view in the center of the GUI, it has fewer functions (for example, not allowing you to preview data).** The terms pane and view have specific meanings in ArcGIS Pro. Panes are dockable windows that contain commands and tools, whereas views contain objects that you are working with, such as a map, 3D scene, layout, or table. A window is a container that can display multiple panes or views. Three windows are visible in figure 1.7: one containing the Contents pane, one containing the Catalog view, and one containing the Catalog pane.

9. **Compare the bottom of the Catalog pane in figure 1.7 with your ArcGIS Pro GUI.** The window in the figure contains three tabs representing three different panes docked in this window: Symbology, Geoprocessing, and Catalog. Your configuration may not match: the Symbology and Geoprocessing tabs are probably not visible in your project at the moment, because you may not have opened the panes yet. One characteristic of the ArcGIS Pro GUI is that it constantly changes depending on what the user is doing.

Objective 1.3: Accessing maps and data from ArcGIS Online

ArcGIS Pro is designed to work closely with ArcGIS Online, so that you can access maps and data that have been saved by others. In the Contents pane, the upper section, Project, lists data stored on the local computer or a local network drive. The lower section, Portal, lists data available in ArcGIS Online or other GIS servers, including your own saved content, My Content; content available to your ArcGIS Online groups, the table in the graphics Groups; all content available in ArcGIS Online, All Portal; and the ArcGIS Living Atlas of the World data available to ArcGIS Online subscribers.

Much of the content accessed by ArcGIS Pro (figure 1.8) is like the content available to ArcMap, although the terminology has changed. A feature service and an image service are now called a *feature layer* and an *imagery layer*, respectively. Next, you can search for and add a couple different types of data.

	Name in ArcGIS Pro	Name in ArcMap	Description
	Feature layers	Feature service	Serve point, line, or polygon features and their attributes
	Imagery	Image service	Serve raster data products such as satellite images or aerial photography in analyzable form.
	Map image layers	Map service	Map cartography based on vector or raster data dynamically rendered and served as image tiles.
	Tile layers	Hosted tile services	A collection of web-accessible tiles that reside on a service, prerendered for rapid display. Includes basemaps and vector basemaps.
	Elevation layers	Elevation services	Cached elevation image tiles in a compressed format, suitable to show terrain in 3D scenes.
	Web tools	Geoprocessing services	Geoprocessing function that runs on the GIS server rather than using local GIS software capabilities.
	Scenes	Scenes	Symbolized 3D spatial content for visualization. May include 2D and 3D layer services.
	Web maps	Web maps	Maps based solely on GIS services. May include other types of services listed in this table.
	Packages	Packages	Data uploaded to ArcGIS Online for sharing. Includes layer, map, project, rule, or geoprocessing packages.
	Scene layers	Scenes	Collection of 3D objects and z-values, including points, 3D objects, integrated meshes, or point clouds.
	Tables	Tables	Rows and columns of information. May contain locations and be drawn on the map.

Figure 1.8. Types of services available to ArcGIS Pro in ArcGIS Online. The terminology has changed somewhat for ArcGIS Pro, and the services are shown with different names than when adding them in ArcMap.

1. **In the Contents pane, click the All Portal entry and type a search term in the box in the Catalog view.**

2. **Note the different types of items available. Click on one of the items to view its metadata.**

3. **Click the Preview option at the bottom of the metadata section to examine a preview. Switch between the Details and Preview tabs as needed.**

4. **Take a moment to examine the Insert tab that is currently visible.**

5. **Right-click a web map ▣ in the Catalog view and click Add and Open.** A web map is a shared map that accesses only GIS services. The map and its contents are displayed in the project. A new map view is added to the center window, and the Contents pane now shows the map layers, much like the ArcMap Table of Contents. In the Catalog pane, a new heading, Maps, is added.

6. **Expand the Maps entry in the Catalog pane and note that the map is now listed there.**

7. **Examine the ribbon along the top.** It switched automatically from the Insert tab to the Map tab when the map was added.

8. **Switch back to the Insert tab.** You may notice that additional commands are now available. Tabs update depending on what the user is doing and what is open.

9. **Return to the Map tab, and examine it for familiar capabilities, such as the basemap, bookmarks, adding data, and so on.** The Catalog view remained open as a second tab in the display window.

10. **Use the tabs to switch back and forth between the map view and the Catalog view.**

11. **Open the Catalog view again, right-click a feature layer 🛡 (not any other kind of layer) from your search group, and click Add to New > Map to add it to a new map, different from the first.**

12. **In the Contents pane, click twice slowly on the map name (currently Map) and change it to something more descriptive.**

13. **Examine the contents of the Maps folder in the Catalog pane, noting that it now has two maps listed.**

14. **Close one of the maps using the X on the tab at the top of the display window.** The map closes but is still available in the Catalog pane.

15. **Click the Save button on the Quick Access Toolbar at the top of the GUI to save the LearnPro project.** The connections to your new maps will be saved to the project.
    Like ArcMap, ArcGIS Pro stores only links to these ArcGIS Online maps and data services.

TIP *Maps in projects are analogous to data frames in an ArcMap map document.*

Objective 1.4: Arranging the windows and panes

As in ArcMap, windows can be docked in the program or float above it. Objects within windows can be rearranged in different ways: stacked on top of each other, side by side, top over bottom, and so on. The docking icon should already be familiar to ArcMap users, and it works much the same way in ArcGIS Pro (figure 1.9).

Figure 1.9. The docking icon is used to arrange panes within a window: stacked (drop on the center) or side by side (left, right, top, or bottom).

1. **In the display window, click the map tab and drag to pull it out of the program onto the desktop, making it a floating window.**

2. **Drag it back into the center display area to make the docking icon appear. Drag the map tab to the center of the icon and release to stack the map back on top of the Catalog view again.**

3. **Drag the map away again and use the docking icon to arrange the map view and Catalog view side by side. Experiment until all five docking locations have been used.**

4. **In the Catalog pane, open the Maps entry, if needed, and double-click the second map to open it.**

5. **Experiment with arranging the three views.** Panes can also be arranged within their windows.

6. **Open the second map, the one containing the feature layer, and click its title bar to ensure that it is the active view.** The Contents pane shows the content of the active view.

7. **If the feature layer loaded as a group layer, expand it until an individual layer is visible.**

8. **In the Contents pane, right-click the layer and click Symbology to open the Symbology pane.**

9. **Drag and dock the Symbology pane in different locations: below the Contents pane, on top of the Catalog pane, next to the Catalog pane, and so on until comfortable with arranging panes.**

10. **Save the LearnPro project.**

TIP *When two panes are stacked in a window, use the top bar to move the window with all the panes together. To move a single pane out of a window, click its tab on the bottom.*

Objective 1.5: Accessing the help

The help files are normally accessed in a web browser window, rather than being part of the program as in ArcMap. For offline work, the help files can also be downloaded and installed separately.

1. **Open the Project ribbon and click Help.** Your default browser opens to the help pages.

2. **Widen the browser window, if needed, to see the outline on the left and the text on the right (figure 1.10).**

3. **Expand some entries in the outline to explore the content and organization.**

4. **Find the Search icon and click to open the search page. Type a search term, such as** labels, **click Search or click Enter on your keyboard, and select one of the topics that appears.**

5. **Examine the breadcrumb trail to locate the content in the Table of Contents outline. Click on a higher level to climb out to more general topics.**

6. **Examine each of the tabs, and then return to the Help tab.**

7. **Spend some time exploring ArcGIS Pro Help until you are comfortable using it.**

8. **Minimize the browser, making the help access quicker next time, and return to ArcGIS Pro.**

Figure 1.10. The help for ArcGIS Pro.

Objective 1.6: Importing a map document

ArcGIS Pro can easily import existing map documents from ArcMap. Every data frame in the map document will be imported as a map, and new layouts will be created in the project. Folder connections needed to access the data in the map document, styles used to display the features, and most data frame and layer settings will be imported. If you have a map document (.mxd) on your computer, you can try importing it now.

1. **Close any open maps.**

2. **Open the Insert tab, and in the Project group, click the Import Map button.**

3. **Navigate to a map document, select it, and click OK. Wait while the map document imports from ArcMap—it may take a few minutes.**

4. **Examine the map views in the main window. One is created for each data frame in the map document.**

5. **In one of the imported maps, double-click a layer to open its properties and examine the Source setting. Close the properties when finished.** The data upon which the map layers are based is imported as paths to the original source in its original location. No data will be copied into the project.

6. **In the Catalog pane, expand the Layouts entry and find the new layout added from the map document. Double-click it to open it.** Most settings will be copied to the new layout, but you may find that a few are not because of differences between ArcMap and ArcGIS Pro. You may need to adjust some layouts or maps to get the same result as the original. Once imported, however, layouts cannot be exported back to ArcMap format.

7. **Examine the layout and see if you notice any difference between the ArcGIS Pro and ArcMap versions.**

8. **In the Catalog pane, expand the Styles entry.** Custom styles used in the map document, if any, would be imported and shown here.

    Remember, ArcGIS Pro can have multiple layouts.

9. **Import another map document if you wish.**

10. **Save the LearnPro project to save the imported map(s) and layout(s).**

Chapter 2

Unpacking the GUI

Background

Users will find many similarities between the ArcMap and ArcGIS Pro GUIs but also some important differences. The ArcGIS Pro GUI is built on the same type of "ribbon" interface common to more recent versions of office software. It is more intuitive but harder to grasp for people accustomed to the ArcMap interface, because the behavior and locations of menus have changed considerably.

The ribbon and tabs

As discussed earlier, the ArcGIS Pro interface contains three types of objects: tabs on the ribbon that contain commands and settings, panes that contain information and settings, and views that contain objects being worked on.

The ArcGIS Pro GUI is highly context sensitive, meaning that it changes on the basis of what the user is currently doing. This arrangement makes the relevant tools more accessible, but it can be confusing because commands and tools do not stay put. Many tabs and commands are hidden until the appropriate context is established. For example, a layer must be selected in the Contents pane (by clicking on it) before the Labeling tab and Symbology pane can be accessed.

The layout and terminology of the ribbon are shown in figure 2.1. Core tabs such as Map and Insert are always visible, whereas contextual tabs appear only when a suitable object is selected. In figure 2.1, a layer has been selected in the Contents pane, causing the Feature Layer contextual tab set and its contextual tabs—Appearance, Labeling, and Data—to appear. Tab functions are organized into groups with titles underneath (Clipboard, Navigate, Layer, and so on). Tools, such as the Explore tool on the Map tab, in the Navigate group, cause the mouse to perform specific functions such as zooming or identifying a feature. Clicking a button takes an action such as zooming in, opening a pane, or opening a tool. Drop-down buttons, when the small inverted triangle is clicked, expand to show additional buttons from which to select (e.g., the Add Data drop-down button in figure 2.1). Finally, the small gray box, or dialog box launcher, on the lower right of some groups (as in the Navigate group in figure 2.1), opens options or settings associated with that group.

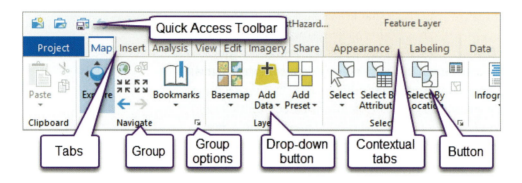

Figure 2.1. Ribbon layout and terminology.

Panes

Panes contain a wide variety of tools, settings, and functions. Many different panes are available. Some, such as the Contents pane and Catalog pane, are needed almost constantly and are nearly always open. Other panes are used occasionally and can be closed when not in use.

   Each pane can itself be organized into many sections, packing myriad functions and settings into a small space. Figure 2.2 shows the different sections available in the Symbology pane. The main pane (figure 2.2a) lets the user choose the map style for a layer, in this case the Lake layer, currently set to the Single Symbol map type. Clicking the symbol opens the Format Polygon Symbol mode (figures 2.2 b and c). The Format Polygon Symbol mode has two tabs: the Gallery tab (figure 2.2b) allows the user to choose from basic symbols, and the Properties tab (figure 2.2c) is used to manipulate the detailed settings of the symbol template. Panes can also possess secondary navigation tabs represented by icons, allowing the user to switch between sets of related functions or settings. For ease of reading, however, this book will adopt the term *secondary tabs* or *tabs* instead of "secondary navigation tabs." The Properties pane, for example, is organized into three different secondary tabs. On each secondary tab, the settings are often organized into groups by headings. The back arrow can be used to return to the main pane. Finally, many panes contain an Options menu button that provides additional functions or settings.

   Although this layout seems complicated, in practice it usually makes sense. The greatest difficulty for ArcMap users is finding a specific setting they are used to. It helps to spend some time exploring and playing with these panes until you learn where the common settings are found. It also helps that tabs on the ribbon often contain the basic settings that are used most frequently. For example, the Labeling tab contains basic settings for labels such as font, size, and color. Users need not employ the more complicated Label Class group unless they must modify details such as halos, rotation, or other advanced settings.

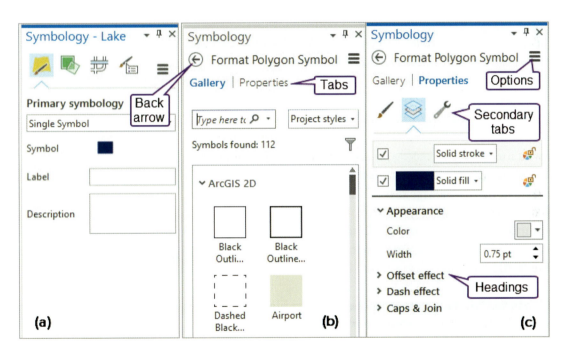

Figure 2.2. The Symbology pane shows how panes can be organized.

Views

   Views are objects that are manipulated by the tabs and panes. The GUI references seven different types of views. Figure 2.3 shows the five view types used most often during mapping and analysis: (1) map views, containing maps and layers (analogous to *data frames* in ArcMap); (2) scene views, containing 3D visualizations; (3) the Catalog view, which permits searching and metadata viewing/editing; (4) a table view, showing the contents of an attribute

table; and (5) a layout view, used to create a map layout for printing. Two additional views, the Fields view and Domains view, are
used to modify the structure of feature classes and geodatabases.

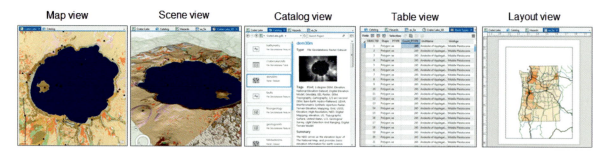

Figure 2.3. The five primary
views in ArcGIS Pro.

Time to explore

In this lesson, you'll explore different aspects and functions of the
GUI using an existing project created for Crater Lake, Oregon.

Objective 2.1: Getting familiar with the Contents pane

Now you'll explore an existing project, and this one has more than
a map. A project named CraterLake that you downloaded and
unzipped earlier is waiting in the SwitchToProData folder.

1. **Open the Project tab and click Open 📑 to switch projects (or
    use the Open button on the Quick Access Toolbar at the top
    of the GUI).**

2. **Navigate to the place you extracted the zip file, and open
    first the SwitchToProData folder and then the CraterLake
    folder. Click and open the project file, CraterLake.aprx
    (figure 2.4). Save the LearnPro project when asked.**

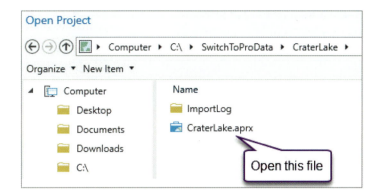

Figure 2.4. Opening the .aprx file to open a project.

The Contents pane is similar to the ArcMap Table of Contents, and users will find many familiar menus and functions. For example, graphical tabs may be selected (figure 2.5) to present the pane in different modes with different functions.

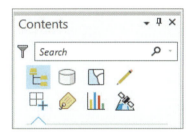

Figure 2.5. Graphical tabs used to present the Contents pane in different modes with different functions (*from upper left to lower right*): Draw Order, Source, Selection, Editing, Snapping, Labeling, Charts, and Perspective Imagery.

3. **Examine the icons at the top of the Contents pane and point to each one to learn its name. Which ones are also found in ArcMap? Which ones are new? Examine each one, and then return to List By Drawing Order.**

4. **Turn layers on and off and rearrange the order.**

5. **Right-click a layer and examine the menu items, noting which are familiar from ArcMap and which appear different or new.**

6. **Open a table for one of the layers, noting that it appears as a view in the display area.**

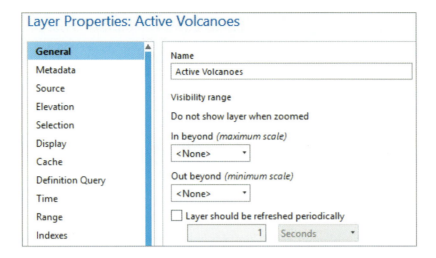

Figure 2.6. Properties in ArcGIS Pro are shown as headings rather than tabs, as shown by these layer properties.

7. **Double-click a layer to open its properties (or right-click a layer and click Properties).** Like ArcMap, the appearance and behaviors of layers and maps are manipulated by settings. As you examine these settings, make note of which ones are familiar, which ones are new, and which ones might be missing.

8. **Examine the available settings, shown in ArcGIS Pro as a list (figure 2.6) rather than as the tabs used in ArcMap. Click each entry in the list and examine the group of settings available for it. Click Cancel to close the properties without changing anything.**

9. **Still in the Contents pane, double-click or right-click the Crater Lake map icon** ![icon] **to open the map properties (analogous to the ArcMap data frame properties). Examine each group of settings and click Cancel when finished to exit without saving any changes.** In ArcGIS Pro, the Contents pane has additional functions and is used to show elements of different types of views.

10. **Open the View ribbon and click Catalog View to open the Catalog view.** Notice that the Contents pane now shows the folders of the project and the ArcGIS Online portal.

11. **Switch back to the Crater Lake map view. The Contents pane now shows the map layers.**

Objective 2.2: Learning to work with objects and tabs

Tabs can get confusing because the same tabs are not always visible, and because the contents of the tabs may change depending on what the user is doing. The key point to remember is that contextual tabs are shown only when the objects they work with are available, and the tab contents are sensitive to the actions that are permitted in the current situation.

1. **In the Contents pane, click the map name, Crater Lake, to highlight it. Examine the ribbon configuration.**

2. **In the Contents pane, click the Vents layer to highlight it (figure 2.7). Notice that new tabs have appeared: a Feature Layer contextual tab set with three contextual tabs: Appearance, Labeling, and Data.**

3. **Click each of the three contextual tabs and examine the settings available. You will learn to use these tabs later.** A layer must be selected to view the contextual tabs that allow the symbols or labels to be changed. This requirement makes sense, because the ribbon must "know" which layer the user wants to change. In ArcMap, this specification was accomplished by opening separate properties for different layers. In ArcGIS Pro, changes in the ribbon apply to the currently selected layer in the Contents pane.

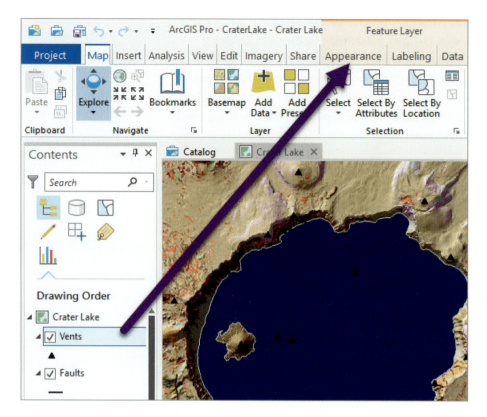

Figure 2.7. The Feature Layer contextual tab set and its contextual tabs appear when a layer is selected.

4. **In the Contents pane, click the Hillshade layer and examine the changes to the ribbon.** As a raster layer, the Hillshade symbolization options are necessarily different. Notice that the new contextual tab set is titled Raster Layer, and there is no Labeling tab.

5. **In the Contents pane, click the Topographic layer, representing the basemap.** Basemaps cannot be resymbolized using these tabs, so even fewer options are available on the tab, now simply called Layer.

TIP *Vector tile basemaps can be resymbolized and customized using a special editor. Consult the help for more information.*

As in ArcMap, there are often multiple ways to access the same function, and you can use whichever one is more familiar or more convenient in context.

6. **Find the three different ways to open the Symbology pane: (1) right-clicking a layer in the Contents pane, (2) clicking the symbol of a layer in the Contents pane, and (3) clicking the Symbology button on the Feature Layer: Appearance tab.** As in ArcMap, the fill color of a symbol can be modified by right-clicking the symbol in the Contents pane and selecting a new color. Try it now if you wish.

   Resist the temptation to start exploring the Symbology settings now. You will tackle this skill later.

TIP *Initially, panes will appear in various places in the GUI. Use your docking skills to arrange them the way you prefer, and ArcGIS Pro will remember these preferences.*

Objective 2.3: Exploring the Catalog pane

The Catalog pane provides a quick way to access the resources available to a project. It has fewer functions than the Catalog view (for example, it can't view metadata), but it is narrow and fits easily in the program window for rapid access.

Data content usually comes from two primary sources: a local drive or internet GIS services.

1. **Examine the Catalog pane. Click the first tab, the Project tab (figure 2.8a), and examine the resources, including Maps, Toolboxes, Folders, and so on.**

TIP *If the Catalog pane is inadvertently closed, open it from the View tab.*

Local resources are stored on the computer's hard drive(s) or network servers mounted on the computer. You'll explore the contents of these folders in the next chapter when we discuss projects. Some folders appear only when the right content is available. No Layouts folder is visible if the project has not yet had any layouts created for it, as is the case for the CraterLake project.

2. **In the Catalog pane, switch to the Portal tab (figure 2.8b). Click each of the graphical secondary tabs in turn, showing different configurations of the pane.**

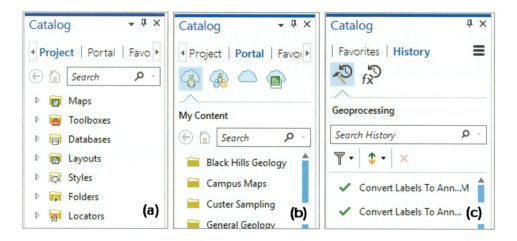

Figure 2.8. The Catalog pane showing (a) the project contents, (b) portal resources, and (c) recent geoprocessing actions.

The Portal tab shows content available over the internet. If you signed into ArcGIS Pro using your ArcGIS Online account, you will see its contents shown. The first tab, My Content, provides access to saved content in your ArcGIS Online account. The second tab, Groups, shows content available in your ArcGIS Online groups. The third tab, All Portal, allows all the ArcGIS Online data resources to be searched, and the final tab, Living Atlas, shows ArcGIS Online subscription content known as the Living Atlas, an authoritative and curated collection of data provided by Esri and its partners. A subscription account is required to access data in the Living Atlas.

3. **In the Catalog pane, switch to the Living Atlas secondary tab ☁, if available. Type an interesting search term, click Enter, and examine the content available.**

4. **Switch to the Catalog view. In the Contents pane, click the Living Atlas entry.**

5. **Type the same search term in the Catalog view and examine the metadata for the items.**

TIP Do not confuse the Catalog view and the Catalog pane. They are similar, but the Catalog view has greater functionality, as shown here in its ability to quickly present the metadata details when searching ArcGIS Online for content.

6. **In the Catalog pane, switch to the Favorites tab.** This tab is probably empty now, but it provides a great spot to store links to often used folders.

7. **On the Favorites tab in the Catalog pane, click the Add Item drop-down arrow and choose Add Folder (or Add Geodatabase). Navigate to a folder (or geodatabase) on your computer or local server where you commonly access GIS data and add it as a favorite.**

8. **Right-click the new Favorites folder connection and note that you can add it to the project now or add it to all new projects in the future.** Note that this Add Item function adds connections only to existing data resources (similar to folder connections in ArcMap).

9. **In the Catalog pane, switch to the History tab (figure 2.8c). (You may need to widen the Catalog pane to see it.)** This tab may be empty if no recent geoprocessing tools have been executed. It keeps track of the tools and the parameters as they are run, similar to the geoprocessing results window in ArcMap. Using this tab, you can click a recent tool and open it, complete with the parameter settings it used.

10. **Save the CraterLake project before continuing.**

Chapter 3

The project

Background

ArcGIS Pro organizes the user experience with data in a completely different way from ArcMap, through a construct known as a *project*.

What is a project?

The project is designed to store all the data, maps, layouts, graphs, tools, and other items that someone uses to do GIS work. Normally, users will create a project that encompasses a specific geographic area and stores the data analyses and products related to it. However, a project may also contain maps or data from multiple regions if a workflow requires them.

TIP *It is a good practice to set the Microsoft® Windows® options for the computer to show the three- or four-letter file extensions— e.g., .docx or .aprx—which help the user understand and identify the file types being used.*

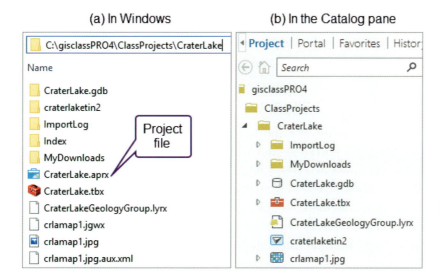

Figure 3.1. Organization of a project as viewed in (a) Windows and (b) the Catalog pane.

Physically, the project is stored on the computer in a folder that has the same name as the project, known as the *project folder*. The CraterLake project shown in figure 3.1a is stored in a Windows folder named CraterLake, which also contains a folder that houses the contents of a file geodatabase of the same name, CraterLake.gdb. The ImportLog and Index folders are also standard parts created for every project, as is the toolbox file (.tbx).

A project folder can also contain objects and folders created by the user because they are associated with that specific project. The craterlaketin2 folder in figure 3.1a contains a three-dimensional TIN surface created using the 3D Analyst extension. The MyDownloads folder was created by the user as a place to collect miscellaneous datasets downloaded during the project compilation phase. Also visible are a layer file (similar to the ArcMap.lyr file but with an .lyrx extension in ArcGIS Pro), and several files associated with a JPEG image named crlamap. Figure 3.1b contains the same project viewed in the Catalog pane. Not all the standard folders are displayed, such as the index, and multifile data objects such as the TIN and the JPEG image are shown as a single entry, just as ArcCatalog does.

TIP *In ArcGIS Pro, it is perfectly okay to store non-GIS files such as Microsoft® Word® documents, spreadsheets, or PDF files in a project folder, and even create subfolders to organize them. As with ArcMap, however, you should never place extraneous files inside the file geodatabase folder, as it risks damaging the geodatabase.*

Items stored in a project

The project is designed to keep track of all the resources needed by the user to effectively work on his or her project goals. Many different items may be included, some of which are familiar from ArcMap, but others are new.

Map. Maps are analogous to ArcMap data frames. They may be created and stored locally within the project or imported from a GIS server web map (such as the one imported to your LearnPro project in chapter 1). A web map is stored as a link back to the original map in ArcGIS Online or through the ArcGIS Portal.

Scene. A three-dimensional visualization is called a *scene*. Originally, scenes were created in the separate ArcScene program in the 3D Analyst extension, but now they are integrated into

ArcGIS Pro. Web scenes are also available from ArcGIS Online and other servers.

Toolbox. ArcMap allowed toolboxes to be created within geo-databases to store custom tools, models, or scripts. In ArcGIS Pro, every project includes a toolbox for storing tools and scripts used for the project.

Geodatabase. Every project includes a geodatabase with the same name as the project and the project folder. It is often called the *project geodatabase* or the *home geodatabase*. The project geodatabase has replaced the cryptic Users location geodatabase as the default location to store outputs from tools. This change greatly benefits novice users who tend to ignore where items are being saved, because it automatically places datasets within the project itself rather than a difficult-to-find user folder. It offers a convenient shortcut for experienced users as well.

Geoprocessing history. As with an ArcMap map document, the project keeps track of every tool run, so that you can review previous steps, check what settings were used to run a tool, or quickly rerun a tool using the same settings or altered settings.

Layout. The layout in ArcGIS Pro is analogous to the ArcMap map document, containing a single-page formal portrayal of one or more maps/scenes for printing or exporting. Projects can contain as many layouts as desired, and the same map can be used in multiple layers and even have different map extents in each layout.

Style. Styles are collections of symbols used to portray map data. ArcGIS Pro comes with default styles (2D and 3D) that are different from the default styles provided with ArcMap, although the ArcMap styles can be imported and used in ArcGIS Pro if desired.

Folder connection. To use data stored outside the project geodatabase, a folder connection to the external data must be established. As in ArcMap, a folder connection can link to a local data folder on the computer, a folder on a local network server, a supported database, or resources on a GIS server. Folder connections are unique to each project rather than being stored en masse as they were in ArcMap and ArcCatalog. The Favorites tab on the Catalog pane, introduced in chapter 2, is a helpful way to quickly add often-used folder connections to a project.

Locator. Locators are used to convert street addresses into points on a map by comparing them with data about streets and their address ranges. ArcGIS Online has a default locator, the World Geocoding Service, which it uses to find a location from a single typed address. Users can create custom locators if they possess the right kind of street data.

Task. Tasks are step-by-step instructions set up to record a certain workflow for future reference, or to share with others to make sure that a procedure is correctly performed each time. Users can record a series of steps as they are performed, and then explain them with detailed instructions. Tasks are a new feature of ArcGIS Pro.

Portal. Portals provide access to GIS services available over the internet. Several types of portals are included: (1) data and services hosted in your ArcGIS Online account, (2) public services hosted by others in ArcGIS Online, and (3) any other services hosted by organizations to which you belong and have access (a login and password may be required).

Project file. The project file, which uses the .aprx extension, is the file in which all these project resources are tracked and managed. It is the file that opens when you open a project, similar to the way a map document was the file opened to begin using ArcMap.

The complexity of the project structure has some implications for data management. Certain techniques commonly employed when using ArcMap may not be available, may not be necessary, or may not work in the same way. You must cultivate different ways of thinking about how you manage data.

TIP When creating a new project, you have the option to decline creating the folder, which merely creates the project file, geodatabase, and other items in the specified folder. From a data organization standpoint, this option is usually not a good choice.

Paths in projects

ArcMap allowed the user to specify whether a map document used absolute or relative paths to refer to datasets. ArcGIS Pro removes this flexibility and enforces a common rationale. Links to data sources on the same hard drive as the project folder are saved as relative paths. Links to data sources on a network drive are saved using absolute paths. Links to data sources on a GIS server are stored using universal locator paths. This method usually provides optimal data handling with minimal problems when projects are copied or moved to new locations. It also allows novice users to safely ignore the absolute/relative path complication, which was never easy to explain or understand.

With this new handling method, links to data are not usually lost when projects are copied or moved (resulting in the dreaded red exclamation points), as long as the entire project folder is copied. Copying a project folder can be easily accomplished using Windows File Explorer (ArcGIS Pro can't do it).

However, deleting, moving, or renaming data folders can still break those path links. Fixing broken paths can be done individually for each layer in a map using the Source section in the layer properties, similar to the method used in ArcMap.

Renaming projects

In short, projects are not designed to be renamed, and it is difficult to do it effectively. Consider the project architecture involved. Perhaps you are unhappy with the name CraterLake, wishing it was CraterLakeOR instead. A glimpse at figure 3.1 demonstrates that several items would need renaming: the project folder, the project geodatabase, the project .aprx file, the project toolbox, and so on. Not only is this procedure a great deal of work, but it is likely all the internal project references to these items would no longer work, with a chance of corrupting the entire project. Unless a "Rename Project tool" becomes available, you should assume that projects cannot be renamed. So make sure you like the name before creating the project!

One might expect that using the Save As command would skirt this issue, but it does not. This command does not create a copy of the entire project folder and its contents. Instead, it creates a copy of only the single .aprx file, with updated links so that the software can find the data. Think of "Save As" in terms of ArcMap files: saving a map document in a new location did not copy all the data to the new file. The data remained in place and was accessed through the folder connections by both the old and new map document (.mxd) files. This procedure prevented users from spawning many copies of the same datasets, which would take up a lot of disk space and result in many different versions of the same information.

However, from experience with word processors and spreadsheets, we are accustomed to thinking that the Save As command creates a complete new copy under a new name, but this assumption is dangerous in ArcGIS Pro. Imagine that an inexperienced Alabama GIS user created a wildlife project named MyProject and filled the project with maps and the geodatabase with datasets. She then decided the project was complete and wanted to share it with her colleagues by putting it on a network drive. She used the Save As command to save it to the network drive with the more descriptive name AlabamaWildlife, and then deleted her MyProject copy. But AlabamaWildlife was only an.aprx file that referred to data saved to MyProject. The user thus lost all her work. She should have instead copied the entire project folder to the network drive although she would still be stuck with the original project name.

Time to explore

These exercises highlight the different types of information stored in a project and where to find them. Continue to use the CraterLake project.

Objective 3.1: Exploring different elements of a project

Project elements are primarily accessed through the Catalog pane or the Catalog view.

1. **Open the CraterLake project.** First, you'll take a tour of the main features on the Project tab of the Catalog pane (figure 3.2).

2. **In the Catalog pane ([1] in figure 3.2), make sure the Project tab is active.**

3. **Expand and collapse the different entries currently visible.**

4. **Collapse all the entries to see everything in the main list.**

5. **Right-click each entry and peruse the context menu that appears.**

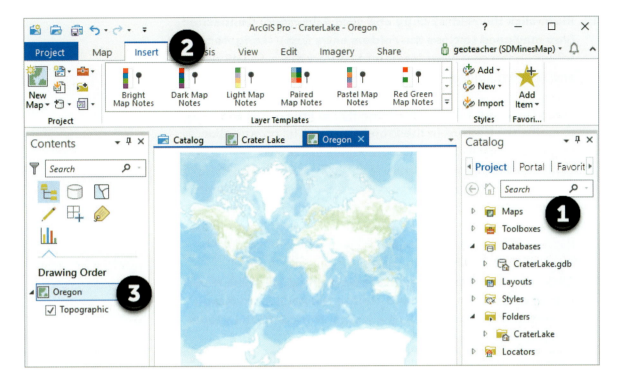

Projects always contain a link or folder connection to the project's home geodatabase. Connections to other databases can be added, and new geodatabases can even be created within the project folder.

Figure 3.2. Using the interface to (1) explore a project, (2) examine the Insert tab on the ribbon, and (3) create a new map named Oregon.

6. **In the Catalog pane, expand the Databases entry to see the project geodatabase.**

7. **Expand the CraterLake.gdb geodatabase to view its contents of feature classes** ⬚ ⬚ ▦ **and rasters** ▦ **.**

8. **Right-click the Databases entry above CraterLake.gdb and click New File Geodatabase (you may need to close the CraterLake geodatabase first). Save it to the SwitchToProData folder and name it ExtraData.**

9. **Expand the Folders entry.** Projects always contain a connection to the project folder. Additional connections can be created to provide access to external data folders outside the project.

10. **Right-click the Folders entry and click Add Folder Connection (you may need to close the Folders entry first). Navigate to a folder on the computer (preferably one containing GIS data) and select it. Click OK.**

11. **Open the Insert tab on the ribbon ([2] in figure 3.2) and examine the buttons in the Project group.** The Insert ribbon has several buttons for adding new objects to a project.

12. **Click the New Map button to create a new map. In Contents, rename the map** Oregon **([3] in figure 3.2).** Some types of items, such as Layouts, have no entry in the Catalog pane because the project does not yet contain any.

13. **Click the Insert tab. Then in the Project group, click the New Layout drop-down arrow, and choose the ANSI – Landscape Letter template to create a new layout. Notice that a Layouts entry appears in the Catalog pane.**

14. **In the Contents pane, change the layout name to** Visit Crater Lake**.**

15. **Expand the Layouts entry in the Catalog pane to see the new layout.**

16. **Close the Visit Crater Lake layout view for now.** You'll learn to create layouts later.

17. **Save the CraterLake project.**

Objective 3.2: Accessing properties of projects, maps, and other items

As in ArcMap, you can set properties that affect how various objects behave. Many of these properties are similar to the settings found in ArcMap. Let's take a tour.

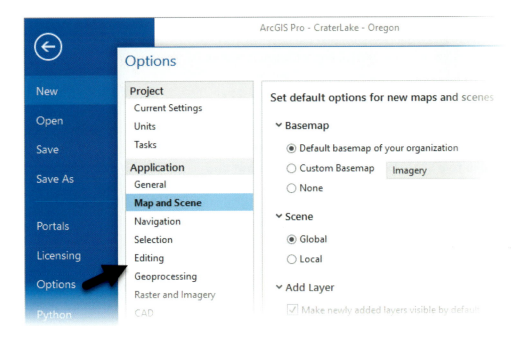

Figure 3.3. Some of the Project options.

1. **Click the Project tab, and then click Options (figure 3.3).**

2. **Examine the list of categories on the left.**

3. **Click each entry in the list and examine the settings that pertain to it.** Take your time, expanding any headings that need to be (marked by a > symbol). Look for familiar settings from ArcMap and note the new ones.

4. **When finished exploring the Project settings, click Cancel to exit without saving any changes.**

5. **Click the large circled back arrow at the top to return to the project.**

6. **Open the Crater Lake map view.** Map properties in ArcGIS Pro are similar to map properties in ArcMap, including how to open them.

7. **In the Contents pane, right-click the icon by the map title** **and click Properties (or double-click the icon).**

8. **Explore the map properties thoroughly and click Cancel when finished.**

9. **Right-click or double-click one of the map layers and thoroughly explore the layer properties.** Layer properties in ArcGIS Pro are also like the ones in ArcMap, with the exception that symbols and labels are now set elsewhere. You won't see these items in the list of settings.

10. **Take note of the Metadata section. In ArcGIS Pro, metadata stored with the source dataset may be copied to the layer metadata. (These two types of metadata are different and may be kept and edited separately.)**

11. **Click Cancel when finished exploring the properties.**

12. **Save the CraterLake project.**

Chapter 4

Navigating and exploring maps

Background

ArcGIS Pro has many of the same functions for exploring and interacting with maps that ArcMap does. Most of these functions are organized on the Map tab (figure 4.1).

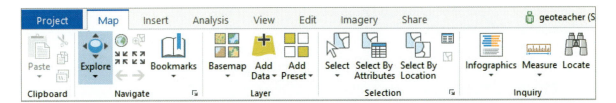

Figure 4.1. Functions on the Map tab primarily explore and interact with the map.

Exploring maps

The Navigate group has many familiar buttons for zooming, but the standard Zoom In/Zoom Out, Pan, and Identify tools are missing, having been replaced by the Explore tool. When this tool is activated, zooming occurs by rolling the mouse wheel, and panning occurs with a left click and drag, eliminating the need to switch between tools to accomplish these common tasks. A click on the map automatically opens a pop-up window with the information pertaining to the clicked feature (the new version of the Identify tool).

The Selection group also looks familiar. The Select tool dropdown arrow opens the standard set of tools for selecting features with rectangles, lines, circles, and so on. The Select By Attributes and Select By Location buttons, however, now open geoprocessing tools instead of opening windows to specify the selection.

The Inquiry group contains updated versions of the Measure and Find tools. The latter is called Locate in ArcGIS Pro. Infographics is an interactive tool to explore the characteristics of a location using the Esri GeoEnrichment service. It can be configured to draw information from a variety of data sources. It requires an ArcGIS Online subscription organizational account and uses service credits.

2D and 3D navigation

ArcGIS Pro integrates the previously separate programs of ArcMap and ArcScene, allowing both 2D maps and 3D scenes to be viewed in the same application without requiring the 3D Analyst extension. The same Explore tool is used for both 2D and 3D navigation.

A 3D scene may be viewed as a global scene or a local scene. A global scene is used to show large regions of the earth's surface in which the curvature of the earth is a component of the scene, such as a scene showing ocean shipping lanes. A local scene is used for smaller regions and can be portrayed in any coordinate system. City views are a common example of local scenes. The viewing method of a scene can be changed from local to global, or vice versa, using the View tab.

Even better, 2D maps and 3D scenes can be viewed side by side, and even linked together. Zooming or panning in one linked scene causes the other to change also, expanding the ability to view and compare multiple visualizations of information.

The Link Views function is found on the View tab (along with buttons for opening other views and panes should they be needed, or accidentally closed). Figure 4.2 shows a 2D map view, *center*, and a 3D scene view docked side by side. They have been linked using the Center and Scale option, which updates the center and zoom level of both views. The Center option updates only the center. Notice that the 2D map view has been rotated so that north is no longer up. After using linked views, it is customary to reorient the map view to north.

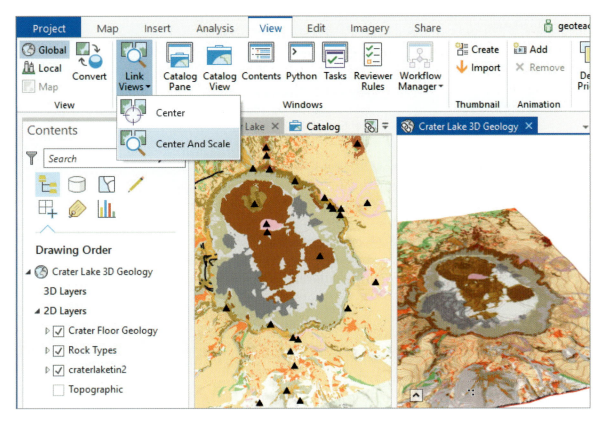

Figure 4.2. Linked 2D and 3D views of Crater Lake geology.

The 3D visualization view requires an elevation dataset upon which the feature classes and other rasters are draped, known as an *elevation source*. The default source is a free ArcGIS Online elevation service named WorldElevation3D/Terrain3D (figure 4.3), so internet service is required to use a 3D view. However, you can use the map properties to specify an alternate or additional elevation source if you have a better one, or if you will be working offline. The order in which the elevation sources are listed in the Contents pane determines the order of precedence, with the top one taking the highest priority.

For example, the Terrain3D service portrays the surface of Crater Lake as flat, which is standard for a lake on a topographic map. However, a raster or TIN containing bathymetric information can be configured to take precedence over Terrain3D by adding the dataset to the Elevation Surface layers in the Contents pane and placing it first on the list. The scene will use the bathymetry where it is available but default to the Terrain3D surface elsewhere.

Figure 4.3. Elevation sources for a scene can be added to the Contents pane.

A 3D scene may include 2D and 3D layers, separated into two different sections of the Contents pane. A 2D layer is a regular map layer that is draped on an elevation surface in a scene. A 3D layer is one that is stored three-dimensionally and has 3D capabilities, such as being displayed with 3D symbols or presented as a billboard with the face always toward the viewer. A 2D layer may be converted to a 3D layer by dragging it to the 3D layer section in the Contents pane.

Time to explore

Objective 4.1: Learning to use the Map tools

The Map tab contains a variety of tools for exploring and interacting with a map (figure 4.4). The Explore tool is used both to navigate and to get information about objects in a map.

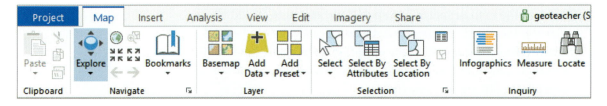

Figure 4.4. Functions on the Map tab primarily explore and interact with the map.

1. **Open the CraterLake project, and make sure the Crater Lake map is open.**

2. **Open the Map tab and point to the Explore tool to see the pop-up window with the button configuration and shortcuts for using the tool. Study them, and then try them out. If you need more information, click F1 while the pop-up window is open to go to the help for the Explore tool.**

3. **Practice until you are comfortable zooming and panning the map.**

4. **Notice that zooming uses preset scales by default. Press and hold the right mouse button while moving the mouse to execute a continuous-scale zoom.**

5. **Click on the map to generate a pop-up window with attribute information for the feature clicked.** Unlike ArcMap, you don't need a different tool to identify features on the map.

6. **Point to it and try the different icons in the pop-up window to see what they do.**

   Bookmarks in ArcGIS Pro work nearly the same way as before but have nice little thumbnails.

7. **Use the Crater Lake bookmark that was already created for this map.**

8. **Zoom in to Wizard Island on the west side of the lake. Create a new bookmark named** Wizard Island**.**

9. **Zoom back to Crater Lake using the bookmark.**

   As in ArcMap, adding data to a map can be done using the Add Data button, or by dragging the dataset from the Catalog pane.

10. **On the Map tab, in the Layer group, click the Add Data button to add the ParkBoundary feature class from the CraterLake project geodatabase. (Then turn the layer off.)**

TIP *Use the Databases or Folders entries at the top of the Add Data browsing window to quickly access the project geodatabase or the project's folder connections (figure 4.5).*

11. **Use the Catalog pane to find the bathymetry feature class in the CraterLake project geodatabase. Drag it to the map.**

12. **Right-click the ParkBoundary and bathymetry layers and click Remove to get rid of them.**

13. Change the basemap to a different one using the Basemap button in the Layer group.

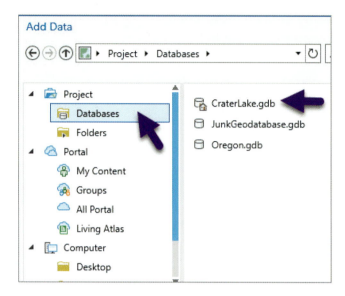

Figure 4.5. Use the Project entries to quickly get data from the project home geodatabase.

The interactive selection tools are also like the ArcMap tools, able to select by rectangles, lines, polygons, and so on.

14. Click the Select drop-down arrow, and experiment with using the tools to select features on the map. Use the familiar Clear button ☒ in the Selection group to clear each selection.

15. Skip the Select By Attributes and Select By Location tools for now. We'll cover them later.

The Measure tool has a new interface but works in much the same way.

16. Click the Measure drop-down arrow and choose to measure Distance, Area, or Features.

17. Examine the instructions and experiment with the options and icons in the tool.

18. To try a different type of measurement, click the Measure drop-down arrow on the Map tab, in the Inquiry group, again, and choose Distance, Area, or Features. Try all three. To close the Measure menu, you must select a different tool.

19. On the Map tab, in the Navigate group, click the Explore tool.

The Locate tool in the Inquiry group is used to find a specific location. It will use the ArcGIS World Geocoding Service unless you specify a different one.

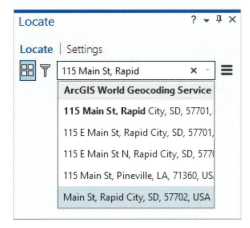

Figure 4.6. Using the Locate pane to search for an address.

20. **Click the Locate tool and look for the Locate pane to appear (figure 4.6). Enter a search string in the Search box, such as your home address. Click Enter to find and zoom to it.**

21. **Find other places if you like.**

22. **When finished, click the Full Extent button ⊕ in the Navigate group.** In ArcMap, the extent used by the Full Extent button defaulted to the extent of the largest layer in the map. In ArcGIS Pro, that largest layer is usually the basemap. Thus, the Full Extent button is normally useless until you set the extent in the map properties.

23. **Use the Crater Lake bookmark to return to the extent of Crater Lake.**

24. **Open the Crater Lake map properties and click the Extent settings.**

25. **Click Custom extent.**

26. **Use the Calculate from drop-down arrow and set it to either Current visible extent or one of the geologic map layers. Click OK.**

27. Zoom way out, and then use the Full Extent button to return to Crater Lake.

28. Save the CraterLake project.

Objective 4.2: Exploring 3D scenes and linking views

In objective 4.1, you learned to pan and zoom in 2D maps using the Explore tool. The same tool for 2D is used to navigate and explore 3D scenes.

1. In the Catalog pane, open the Maps entry and double-click the Crater Lake _3D scene to open it (figure 4.7).

TIP Maps and scenes must be double-clicked to open them. They cannot be dragged to the display window.

2. Open the Map tab. Point to the Explore tool again and study the pop-up window to review the 3D navigation techniques.

3. Click the View tab and examine the View group. A scene can be viewed as a local scene or a global scene. Make sure that the Crater Lake_3D scene is set to Local.

4. Practice zooming, panning, tilting, and rotating the 3D scene.

Now you'll learn how to link views so that they respond together.

5. Dock the Crater Lake_3D view on the side of the display window, and select the Crater Lake map view, if needed, so that the 2D map and the 3D scene are side by side, as shown in figure 4.2.

6. Do some navigation in both views separately for a moment.

7. Open the View tab, and in the Link group, click the Link Views > Center and Scale option.

8. Navigate in one view or the other and watch the linked view update.

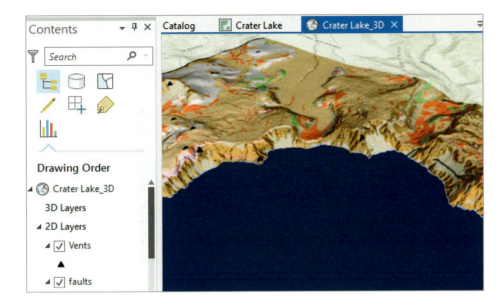

Figure 4.7. A 3D scene showing part of Crater Lake.

Notice that the Contents pane updates depending on which item is currently the active view in the display window.

9. **Click on the Crater Lake view tab to make it the active view (the tab will appear blue) and examine the Contents pane.**

10. **Click on the Crater Lake_3D view tab to activate it and examine how the Contents pane changes.**

TIP Maps can also be linked to maps, and scenes can be linked to scenes. You can even link more than two views together, although display performance may begin to suffer if many links are in use.

Links can be removed once you no longer need them. However, the previous navigation has rotated the Crater Lake map view so that north is no longer up. It is easiest to fix this issue before removing the link.

11. **In the Crater Lake_3D scene, click the north arrow on the Navigator compass ring to reorient both the scene and map.**

12. **Switch to the View tab and click the Link Views button again to turn off the link (its color will change from light blue to white).**

13. **Dock the Crater Lake_3D scene on top of the Crater Lake map so that only the scene is visible.**

14. **Turn off the Lake layer, if needed, and zoom/rotate to examine the geologic units inside the crater.** Notice that the geologic units inside the crater appear flat at the level of the lake. The Terrain3D service used for the 3D visualization portrays the lake area as flat. A bathymetric dataset can be configured to take precedence over the Terrain3D elevation source where it is available.

15. **In the Contents pane, scroll to the bottom and examine the Elevation Surfaces section showing the current surface, Terrain3D.**

16. **Right-click the Ground entry and click Add Elevation Source.**

17. **Navigate to the CraterLake project folder (not the geodatabase) and find the craterlaketin2 TIN dataset. Add it.**

18. **Make sure the TIN appears at the top of the Elevation sources list.**

19. **Zoom and pan to examine the crater bathymetry.**

20. **Close the Crater Lake_3D scene.**

21. **Save the CraterLake project.**

TIP *If you forget to reorient the map before unlinking, the map rotation can also be reset to 0° in the map properties (figure 4.8).*

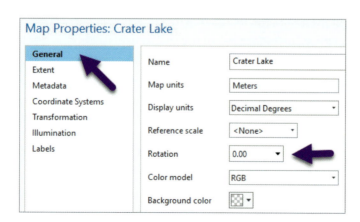

Figure 4.8. Setting the map rotation angle in the map properties.

Chapter 5

Symbolizing maps

Background

ArcGIS Pro uses the same map symbol types available in ArcMap, such as graduated symbols, graduated colors, and so on. However, the interface through which they are selected and manipulated has changed significantly. Instead of being set through the layer properties, symbols are set using the Symbology pane.

Accessing the symbol settings for layers

The Symbology pane is not associated with any particular layer. Instead, the settings impact the currently selected layer in the Contents pane. This arrangement makes it much quicker to switch from one layer's symbol settings to another's: instead of closing and opening a layer's properties, you simply click the layer in the Contents pane.

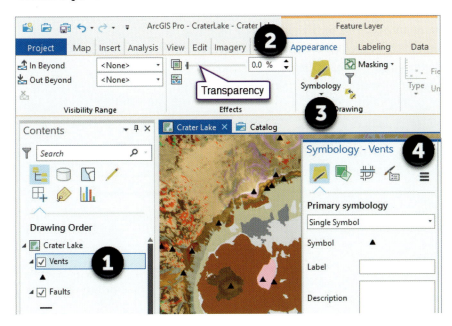

Figure 5.1. Choosing a map and symbols for a layer.

To view the symbology settings first requires that a layer be selected in the Contents pane ([1] in figure 5.1). This selection causes the Feature Layer contextual tab set to appear. Under the Feature Layer contextual tab are three contextual tabs: Appearance, Labeling, and Data.

Opening the Appearance contextual tab ([2] in figure 5.1) accesses settings for visibility scale ranges, transparency, swiping, masking, and other functions. The Symbology drop-down arrow ([3] in figure 5.1) is used to select the type of map symbology to apply to the layer: Single Symbol, Graduated Symbols, and so on. After a map type has been selected, the Symbology pane opens automatically ([4] in figure 5.1) in case the user wants to modify the symbols, such as changing the black triangle for the vents to some other symbol.

Accessing the labeling properties

The Labeling contextual tab (figure 5.2) behaves in much the same way as the Appearance tab. The settings affect whichever layer is currently selected in the Contents pane. The tab itself provides access to the most common labeling functions: turning them on, choosing fonts and styles, setting visibility ranges, choosing basic placement styles, and setting different classes of labels based on an SQL expression. All these functions were available in ArcMap.

The labeling engine for ArcGIS Pro is more sophisticated than the one provided in ArcMap. In fact, you can use the flexible Maplex engine, which was an additional extension for ArcMap, as the default labeling engine in ArcGIS Pro. The improved labeling means that you can obtain more labeling effects, but it requires manipulating and understanding more settings.

Figure 5.2. The Labeling contextual tab controls the most common label settings.

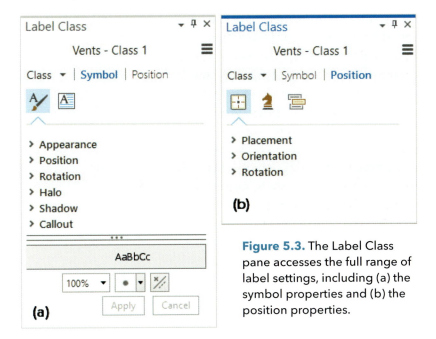

Figure 5.3. The Label Class pane accesses the full range of label settings, including (a) the symbol properties and (b) the position properties.

If the options provided on the Labeling contextual tab are insufficient, clicking the Options dialog box launcher on the lower right of the Text Symbol or Label Placement groups will open the Label Class pane (figure 5.3), which provides access to the many detailed settings. This pane is one of the more complicated ones, with sections for choosing the label class being modified, as well as the label symbol settings (figure 5.3a) and the label position settings (figure 5.3b).

Because labeling has more options in ArcGIS Pro, and perhaps because a map can now be placed in multiple layouts, the ability to create annotation from labels and store it in the data frame (map) is not available in ArcGIS Pro. Annotation must now be created and stored within a geodatabase, and it must be edited using the regular editing tools. Both regular and feature-linked annotation are still available.

ArcMap had a window called the Label Manager that simplified manipulating multiple labeled layers in a map. ArcGIS Pro already facilitates this process because you simply select a different layer to access the label settings on the ribbon, rather than closing and opening properties for each layer. ArcGIS Pro also has a Labeling tab in the Contents pane (figure 5.4) to speed turning labels on and off. The order of the list of entries also sets the priority for labels, with the layers on top taking precedence.

Figure 5.4. The Labeling tab in the Contents pane.

Symbolizing rasters

The options for symbolizing rasters in ArcGIS Pro follow the same ideas and methods as in ArcMap, although, as with other symbols, the interface for setting them has changed. As with symbolizing layers in ArcGIS Pro, the layer must first be selected in the Contents pane to make the Raster Layer contextual tab set appear with the Appearance contextual tab (figure 5.5).

Some of the basic settings in the Rendering group include the Symbology raster display type (Stretch, Discrete, Classify, and so on); Stretch Type, used for stretched rasters; and Band Combination, for multiband imagery. The Band Combination button has two default settings, Natural Color and Color Infrared, that can be used for standard aerial imagery, but custom settings can also be created. In figure 5.5, the Landsat Natural Color setting was created by the user to accommodate the different band numbers used for Landsat and most standard aerial color imagery.

The DRA button applies dynamic range adjustment, which uses statistics calculated from the current display window for stretches. It is helpful when surveying a local area from an image with a much larger extent, providing better display results, although it is somewhat slower. Slider adjustments for brightness, contrast, and gamma are also available.

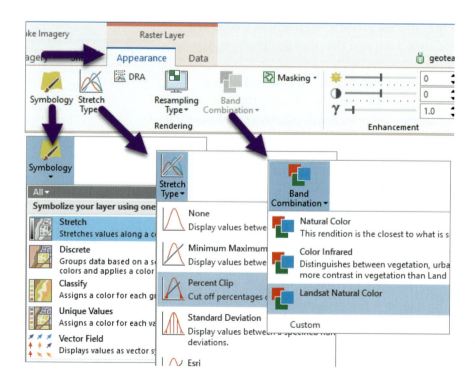

Figure 5.5. Raster symbol settings.

In addition to raster symbology, ArcGIS Pro also has an Imagery tab (figure 5.6) that allows manipulation and display of raster data and is especially useful for multiband imagery such as aerial photography or Landsat imagery. You can quickly display an image using the normalized difference vegetation index (NDVI) or one of the other common indexes without having to create a new image. The ratio is applied on the fly to the display. The Imagery tab also contains tools for classifying, georeferencing, and orthorectifying imagery. We will not try the tools in this book, but if you are familiar with manipulating multiband images, this tab will be a welcome ally.

Figure 5.6. The Imagery tab offers sophisticated tools for image manipulation.

Time to explore

In this section, you will become familiar with how ArcGIS Pro manages symbols, labels, and raster display.

Objective 5.1: Modifying single symbols

To start exploring symbols, you'll return to the Oregon map you created in chapter 2 and add some additional datasets from a feature dataset named Oregon in the CraterLake project geodatabase.

1. **Open the CraterLake project.**

2. **Close all the current views in the center of the window.**

3. **In the Catalog pane, expand the Maps entry and double-click the Oregon map. (If you don't see it, use the Insert tab on the ribbon, and the Project group, to create a new map and name it Oregon.)**

4. **In the Catalog pane, expand the CraterLake project geodatabase. Click the Oregon feature dataset and drag it to the map to add all its feature classes.**

5. **Hold the Ctrl key and click to clear one of the visibility check boxes to turn all the layers off.**

6. **In the Contents pane, turn the volcanoes layer and the Topographic basemap back on.**

7. **Select the volcanoes layer by clicking it.** Remember, the standard way to change a layer's symbols starts by selecting the layer.

8. **Look for the Feature Layer contextual tab on the ribbon, and click the Appearance tab.**

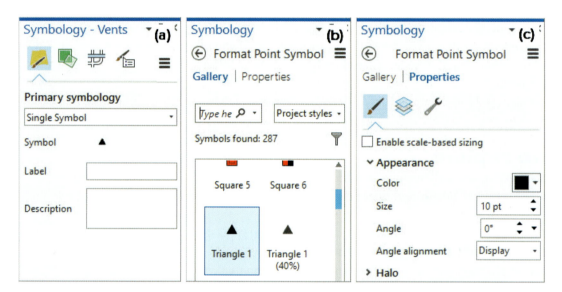

9. **Click the Symbology button to open the Symbology pane (figure 5.7a). Arrange placement of the pane as desired; in the same window as the Catalog pane is a convenient place.**

Figure 5.7.
Modifying a single symbol.

10. **In the Symbology pane, click the symbol representation to change the symbol. It opens the Format Point Symbol mode of the pane.** Notice that the Format Point Symbol mode has two tabs: Gallery and Properties. The Gallery tab (figure 5.7b) is used to select the base symbol, and the Properties tab (figure 5.7c) is used to modify it.

11. **Scroll down and select the Triangle 1 (solid black) symbol. Notice that the symbol changes on the map as soon as you click it.**

12. **Switch to the Properties tab and examine it.** Note that the Properties tab has three different graphical secondary tabs indicated by the icons: Symbol (paintbrush), Layers (stacked layers), and Structure (wrench).

13. **Choose each one of these tabs in turn and examine the settings.**

14. **Return to the first Symbology graphical tab** ✏. This tab contains the commonly used settings.

15. **Expand the Appearance heading. Change the color to reddish brown and the size to 9 pt. Click Apply to make the change take effect.**

16. **In the Contents pane, turn on the airports layer and click it to highlight it. The Symbology pane immediately updates to show settings for the airports layer.** An advantage of the Symbology pane is that you can quickly switch to, and edit, symbols of a different layer.

17. **In the Symbology pane, click the symbol and open the Gallery tab. Type** airplane **in the Search box and click Enter.**

18. **To see more airplane symbols, click the Project styles drop-down arrow and change it to All styles.**

19. **Choose a symbol. Switch to the Properties tab and update the color, size, or rotation.** Setting single symbols for lines and polygons is similar. Now you can practice.

20. **Turn on the other map layers. Select each layer in turn and give it a suitable symbol. Be sure to examine the variety of settings for lines and polygons.**

21. **Close the Symbology pane.**

Many of the familiar symbol-setting shortcuts from ArcMap also work in ArcGIS Pro.

22. **Right-click a layer name and click Symbology to open the pane again.**

23. In the Contents pane, click the symbol of a layer to open the Format Symbol mode.

24. Right-click the symbol of a layer to quickly change its fill color.

TIP When using the Symbology pane frequently, it can be convenient to keep it open and stacked in the same window as the Catalog pane, using the tabs to switch to it when needed.

Objective 5.2: Creating maps from attributes

The Unique Values map in ArcGIS Pro is like the one in ArcMap.

1. In the Contents pane, select the parks layer.

2. On the ribbon, on the Feature Layer: Appearance tab, in the Drawing group, click the Symbology drop-down arrow and choose Unique Values for the map type. (Or choose the map type using the Primary Symbology drop-down arrow in the Symbology pane.)

3. For Field 1, choose FCC for the style to base the map on (figure 5.8).

4. Choose a color scheme.

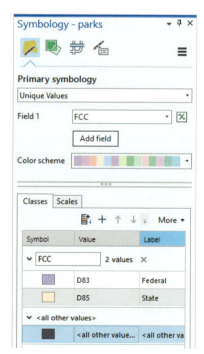

Figure 5.8. Creating a Unique Values map for parks.

5. **Under Classes, change the label for D83 to** Federal. **Change the label for D85 to** State. **(It can be helpful to widen the Symbology pane to see the Label tab.)**

6. **Click one of the color symbols to modify its properties. Then click Apply. Use the back arrow in the Format Polygon Symbol mode to return to the Primary symbology mode.**

7. **On the Classes tab, click the More drop-down arrow and examine each of the other settings available.** Most of them match the settings in the ArcMap Unique Values map menu.

8. **Use the More drop-down arrow to turn off the Show all other values class.**

9. **Click the menu button ☰ in the upper-right corner. Examine the additional settings here, such as importing symbology from a layer or layer file.**

    ArcGIS Pro has a technique known as "scale-based symbology" that allows the user to choose different symbols for different map scales. For example, you might use the classic double-line filled Interstate symbol at larger scales but a plain symbol at smaller scales. Read more about scale-based symbols in the help if you are interested.

10. **Click the Scales tab to switch to the scale range options and examine them.**

11. **Switch back to the Classes tab.** Now you can examine the settings for a Graduated Color map.

12. **In Contents, click the counties layer to highlight it. On the Appearance tab on the ribbon, or in the Symbology pane, choose the Graduated Colors map type.** These settings should look familiar.

13. **Choose the POP2014 field and experiment with the settings.**

14. **Use More > Format all symbols to change the outline color of all classes.**

15. Use More > Show statistics to show the statistical values of the field.

16. Click the menu button ≡ and examine the choices.

17. At the top of the Symbology pane, click the Advanced symbol options graphical tab 📝**.**

18. Expand the Format labels heading and give the labels 3 **significant digits with** thousands **separators.**

19. Click the Primary symbology graphical tab ✏️ **to return to the previous tab.**

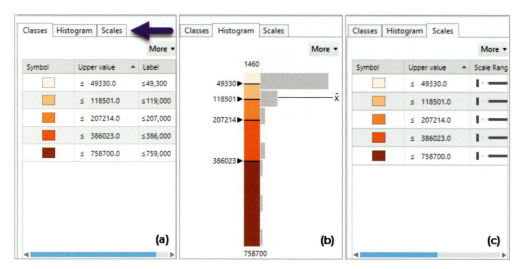

(a)   (b)   (c)

The Primary symbology tab has multiple tabs on the bottom that are handy for managing class breaks. The Classes tab shows the classes, ranges, and labels (figure 5.9a). The Histogram tab (figure 5.9b) shows a histogram of the class breaks. Pointing to the X shows a pop-up window with the mean value. The class break values can be dragged up or down, or the values can be edited, to apply specific manual breaks. The Scales tab (figure 5.9c) allows different symbols to be set for different scale ranges.

Figure 5.9. Tabs for managing different aspects of the class breaks: (a) the Classes tab, (b) the Histogram tab, and (c) the Scales tab.

20. **In the lower half of the Primary symbology tab, switch to the Histogram tab and practice manipulating the class breaks.**

21. **In the Symbology pane, change the map type to Graduated Symbols and experiment with manipulating the symbols.**

22. **Spend 10 or 15 minutes exploring all the map types and settings for different types of maps and geometries (point, line, polygon) until you feel comfortable with the process.** ArcGIS Pro also has some new map types in addition to the familiar ones.

23. **Save the CraterLake project.**

Objective 5.3: Creating labels

The labeling engine in ArcGIS Pro is more sophisticated than the one in ArcMap, and you can use the Maplex extension by default. With more sophistication comes more options, but many basic tasks can be easily handled using the ribbon.

1. **In Contents, click the airports layer to select it.**

2. **Open the Feature Layer: Labeling tab on the ribbon.**

3. **In the Label Class group, make sure Field is set to NAME.**

4. **Click the Label button to turn on the labels.**

5. **Experiment with the Text Symbol group buttons to modify the font, size, or style.**

6. **In Contents, select the rivers layer.**

7. **Zoom in to about 1:400,000 scale.**

8. **Open the Labeling tab on the ribbon (figure 5.10a).**

9. **In the Label Placement group, scroll or click the drop-down arrow in the Label Placement box (if your GUI is not sized wide, it will appear as a Label Placement Style drop-down button), and examine the options (figure 5.10b). Choose Water (Line).**

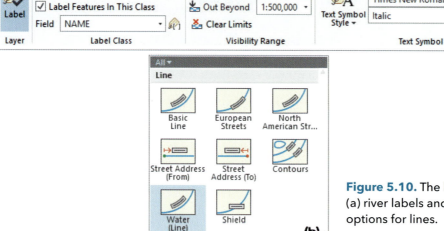

Figure 5.10. The label settings for (a) river labels and (b) the placement options for lines.

10. **Turn on the labels. Change the font to blue italic. (Note: The default Tahoma font has no italic version. Choose a different font first.)**

11. **Set the Out Beyond visibility range to** 1:500,000 **(figure 5.10a) and zoom out/in to verify that it works.**

12. **Zoom to exactly** 1:500,000 **and click the View Unplaced button** **in the Label Placement group to view unplaced labels.**

    Label classes are helpful for labeling different groups of features with different labels. Next, you can give the interstates an appropriate symbol.

13. **In Contents, select the highways layer and turn on the labels.**

14. **On the Labeling tab on the ribbon, set the Label Class field to HWY_SYMBOL.**

15. **For Label Placement Style, choose Shield.**

16. **For Text Symbol Style, choose Shield 1, the familiar red-and-blue interstate symbol.**

17. **On the Labeling tab on the ribbon, click the SQL Query button 🔲 next to the Class box, currently containing Class 1. The Label Class pane opens (figure 5.11).**

18. **In the Label Class pane, click the SQL expression box if needed.**

19. **Click the Add Clause button and enter an expression: HWY_TYPE is Equal to I. (No quotation marks are needed.) Click Add and then Apply.**

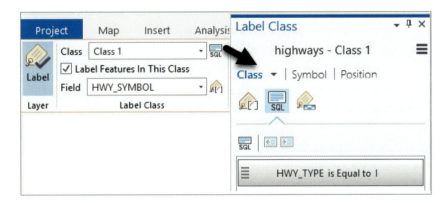

Figure 5.11. Creating a label class for highways.

Now you can create a second class of labels for highways.

20. **On the Labeling tab on the ribbon, click the Class drop-down arrow, and then click Create label class.**

21. **Enter Highways for the label class name and click OK. (If you get a Save Where Clause, it means that you forgot to click Apply in step 19. Click OK.)**

22. **Click the Add Clause button again and enter the expression: HWY_TYPE does Not Equal I.**

23. **Set the label Field to HWY_SYMBOL, choose Shield for Label Placement Style, and Shield 7 for Text Symbol Style.**

24. **Apply a scale range of** 1:1,000,000 **to the Highways label class to reduce the clutter.**

Objective 5.4: Managing labels

ArcMap had a window called Label Manager that simplified manipulating multiple labeled layers in a map. ArcGIS Pro already streamlines this process because you simply select a different layer to access the label settings on the ribbon, rather than closing and opening properties for each layer. ArcGIS Pro has a List By Labeling tab in the Contents pane to facilitate turning labels on and off. The order of the list also sets the priority for labels, with the layers on top taking precedence. You can use it to rename the first label class "Interstates" instead of Class1.

1. **In Contents, click the List By Labeling graphical tab** .

2. **Click the Class 1 heading below the highways layer and edit it to read** Interstates.

3. **Drag the highways layer above the rivers layer so that the highways labels take precedence in case of a conflict.**

4. **Use the List By Labeling tab to make any final adjustments to the highways or rivers labels by clicking the desired class and adjusting the tab settings.**

5. **Create labels for the counties feature class and symbolize them to your satisfaction using the Labeling tab on the ribbon.**

   You can do a lot with the Labeling tab on the ribbon, but sometimes more advanced settings are needed. The Options dialog box launcher in the lower-right corner of the Text Symbol and Label Placement groups accesses more detailed format options.

6. **On the Labeling tab, click the Options dialog box launcher** ⌐ **in the Text Symbol group to open the Label Class pane.**

7. **Examine the three tabs: Class, Symbol, and Position.**

8. **Under each heading, examine the graphical secondary tabs and explore the settings.** It takes a while to master the different settings, remember how to apply them, and learn how they impact one another. For today, you can create some halos for the county labels.

9. **In the Label Class pane, click the Symbol tab and the General graphical tab ✌.**

10. **Expand the Halo entry.**

11. **Click the Halo symbol drop-down arrow and select one of the symbols, such as White fill 50% transparency.**

12. **Change the color if you wish and set the Halo size to 2 pt. Click Apply.**

Maplex is the default labeling engine in ArcGIS Pro, but you can switch to the Standard Label Engine if desired. Any advanced settings made prior to the switch will be lost and cannot be recovered.

13. **On the Labeling tab on the ribbon, examine the three buttons in the Map group.** The top button, Pause, is useful if you accidentally turned on too many labels and want to stop drawing them.

   The middle button, View Unplaced, shows the unplaced labels.
   The bottom button, Labeling Options, controls weights and other advanced features, and allows the label engine to be changed.
   As with ArcMap, label options can also be accessed elsewhere.

14. **In Contents, right-click the airports layer and look for the Label entry in the menu, used to turn labels on or off. Also notice the Labeling Properties entry.**

15. **Spend 10 or 15 more minutes exploring the menus, creating labels and modifying their settings, to get more comfortable with labeling.**

16. **Save the CraterLake project.**

Objective 5.5: Symbolizing rasters

The methods for displaying raster data will be familiar to you from ArcMap, although the way they are implemented has changed. Nevertheless, the ArcGIS Pro method should be starting to feel familiar, too.

1. **Close the Oregon map and use the Catalog pane Maps entry to double-click and open the Crater Lake Imagery map.**

2. **Add the dem30m raster from the CraterLake.gdb project geodatabase.**

3. **Turn off all layers except dem30m, a digital elevation model.**

4. **In the Contents pane, select the dem30m layer and open the Raster Layer: Appearance contextual tab on the ribbon.**

5. **In the Rendering group, click the Symbology drop-down arrow and examine the display choices.**

   The display types are the same as in ArcMap, except for a new Vector Field raster map type. The software preselected the Stretch type based on the characteristics of the DEM raster file.

6. **Click the Symbology drop-down arrow again and choose the Stretch type to open the Symbology pane (or right-click dem30m in Contents and open it that way).**

7. **Select a color scheme and experiment with the Stretch settings and other options in the pane (figure 5.12a). They will be familiar to you if you display stretched rasters in ArcMap.**

8. **In the Symbology pane, change the Stretch map type to Classify (figure 5.12b). Set a color scheme, choose the Equal Interval classification method, and increase the number of classes to 15 or 20.**

9. **In Contents, turn on the Topographic basemap.**

10. **On the Raster Layer: Appearance tab on the ribbon, use the Layer Transparency slider or the % box to set the transparency of dem30m to ~60%.**

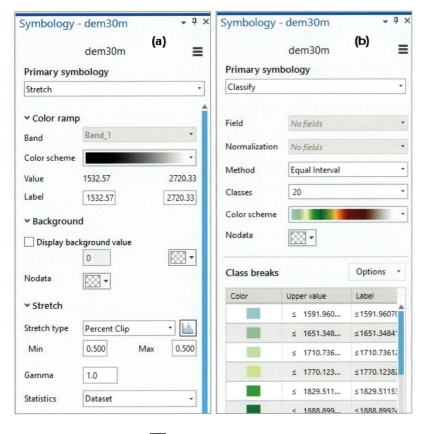

Figure 5.12. Raster symbology showing (a) the stretched display type and (b) the classified display type.

11. **Click the Swipe tool ⬚ . Drag from an edge of the map to peel off the DEM and see the basemap underneath.** You'll select a different tool to stop swiping.

12. **Click the Map tab on the ribbon, and in the Navigate group, click the Explore tool.**

13. **Switch back to the Raster Layer: Appearance tab.**

14. **Turn off all the raster layers except the Landsat image, LT05_Sep11_2011_P45R30clip.tif.**

15. **In the Contents pane, click the Landsat image layer to select it and view the Symbology pane.** Multiband images are still displayed using the red-green-blue (RGB) composite method but with the option of preserving custom settings.

16. **On the Appearance tab on the ribbon, in the Rendering group, click the Band Combination drop-down arrow and choose Custom.**

17. **Name the combination** Landsat False Color**, and set the red to Band_4, the green to Band_3, and the blue to Band_1. Click Add.**

18. **In the Rendering group on the Appearance tab, click the Stretch Type drop-down arrow and try several different stretches. Leave it set to the one you like best.**

Finally, ArcGIS Pro has integrated several functions from the Imagery toolbar in ArcMap, such as quickly creating an NDVI, or greenness index map, using multiband imagery. The index is applied to each pixel dynamically, with no need to create a new raster dataset.

19. **Switch to the Imagery tab on the ribbon (only visible when a raster is selected in the Contents pane).**

20. **Note the various buttons and tools on the tab, including those for georeferencing and classification.**

21. **In the Tools group, click the Indices drop-down arrow and choose NDVI.**

22. **Set Near Infrared Band Index to** 4 **and Red Band Index to** 3**. Click OK.**

23. **A new layer appears showing vegetated areas as bright and nonvegetated areas as dark.**

24. **Spend 10 or 15 minutes experimenting with these different raster display options on both the Appearance and Imagery tabs.**

25. **Save the CraterLake project.**

TIP Layers derived from the Index tools are temporary and may not be available after the project has been closed and reopened. They can easily be removed and regenerated, however, or you can save them permanently by right-clicking the NDVI layer, and then clicking Data > Export Raster.

Chapter 6

Geoprocessing

Background

Geoprocessing in ArcGIS Pro is much like geoprocessing in ArcMap. ArcToolbox is retained, although with a different look, and most of the tools are configured and run in the same way.

What's different

Five significant differences can be mentioned. First, the collection of tools has been impacted by the design decision that most operations in ArcGIS Pro should run as tools, to comply with the multithreaded nature of the program. Some familiar menu tasks, such as selecting records from a table, have been replaced by tools. The right-click may still exist, but now a tool opens instead of a context menu or window.

Second, the somewhat clunky Search window and ArcToolbox have been combined into a single entity, the Geoprocessing pane (figure 6.1a). Searching for a tool is far quicker and easier, and common and recently used tools are remembered. Under the Recent heading (look under the Intersect tool in figure 6.1a), the drop-down arrow for a tool can be used to reopen a recently run tool complete with its set parameters. Most users will find little need, anymore, to open the Toolbox itself or to know in which toolset a tool resides.

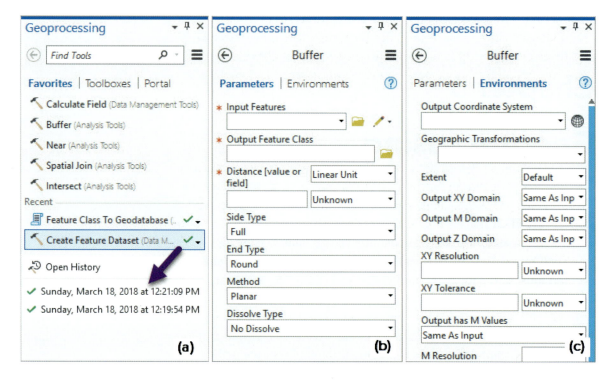

Figure 6.1. Tabs of the Geoprocessing pane including: (a) the Favorites tab, (b) the Parameters tab, and (c) the Environments tab.

Third, the Environments settings, although they contain many of the same items and work in the same way, have become smarter. Each tool has been given an Environments tab in addition to a Parameters tab. Figure 6.1b shows the Parameters tab of the Buffer tool, where the tool settings are entered, and the Environments tab (figure 6.1c), where the Environments settings may be adjusted. Only the Environments settings related to the specific tool are shown, making it easier to quickly view all the relevant settings and make sure they are adjusted as desired. The full panoply of Environments settings can still be accessed from the Analysis tab on the ribbon.

Fourth, the default location for geoprocessing results is no longer the nearly invisible and unsatisfactory default geodatabase in a subfolder of User documents. Instead, outputs are saved by default to the project geodatabase in the project folder. This change is a great improvement because it places all the output files in a logical place where the user can easily find them again. Beginning users especially, who have not yet learned to eschew ArcMap's default storage location, will find this treatment more convenient and logical, and they will be less likely to lose track of data. An alternate default workspace may still be specified in the Environments settings, but this practice has changed from a near necessity to an occasional adjustment to accommodate a specific need.

Finally, the background geoprocessing function of ArcMap is no longer visible. This option has been superseded by the multi-threaded nature of ArcGIS Pro, in the sense that all tools essentially run in the background, quickly and reliably.

Analysis buttons and tools

The Analysis tab on the ribbon organizes access to the geoprocessing functions (figure 6.2). In the Geoprocessing group, the History button opens the corresponding tab in the Catalog pane to review previous processing steps and their parameters. The Python and ModelBuilder buttons open those windows familiar from ArcMap. The Environments button opens the Environments settings, and Tools opens the Geoprocessing pane. The Ready To Use Tools button in the Geoprocessing group and the Feature Analysis and Raster Analysis buttons in the Portal group open geoprocessing services from ArcGIS Online (many of which consume service credits, and can usually be accomplished with the geoprocessing tools in ArcGIS Pro). The remainder of the toolbar accesses specific groups of tools.

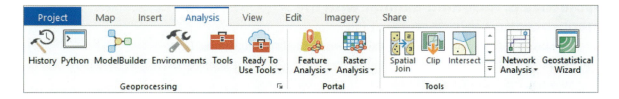

Figure 6.2. A portion of the Analysis tab.

Tool licensing

Tool licensing follows the standard ArcGIS model. Certain tools and toolsets require a purchased extension, such as the Contour tool licensed for use in Spatial Analyst or 3D Analyst. The user must have a license for certain tools to be able to execute them. Information about the extension required can be found in ArcGIS Pro Help. Instructors (and everyone else) will be glad to learn that extensions need no longer be "turned on" to run the extension tools; if the tool is licensed, it will run.

Most users will find that searching for and executing tools is greatly streamlined in the Geoprocessing pane and represents a significant improvement over ArcMap geoprocessing. One great advantage is that a tool remains open after execution, with the parameters intact. Thus, if you must rerun the same process several times, changing only one or two parameters, the sequence can be done more quickly. However, in some cases, it works better to clear the tool by clicking the back arrow to return from the tool (figure 6.1b) to the main Geoprocessing pane (figure 6.1a). This step must be taken when the tool settings are derived from one of the tool parameters. For example, in a selection tool, the fields in the selection expression must be determined from the layer specified at the top of the tool. Changing the layer invalidates the field settings lower in the tool and generates warnings because the fields can no longer be found. In such cases, exiting the tool dialog box (using the back arrow shown in figure 6.1) and reopening the tool from the list is faster and less confusing.

Time to explore

In this section, you'll first explore the new way to execute geoprocessing tools, and then look at some common geoprocessing tasks that will now look and feel somewhat different from ArcMap.

Objective 6.1: Getting familiar with the geoprocessing interface

The layout of the geoprocessing tools is somewhat new, but most users will find an improved experience in many aspects of geoprocessing. You can experiment by running the Buffer tool.

1. **Open the CraterLake project.**

2. **Open the Oregon map and close any other open maps.**

3. **Open the Analysis tab on the ribbon. Examine the buttons and view the contents of each drop-down arrow without selecting any of them.**

4. **In the Geoprocessing group, click the Environments button and examine the new layout of the settings and their defaults. Note the Current Workspace and Scratch Workspace settings (the project geodatabase).**

5. **Click Cancel to close the Environments settings without making changes.**

6. **Click the Tools button to open the Geoprocessing pane.**

TIP It is usually convenient to dock the Geoprocessing pane on top of the Catalog pane, making it easy to switch between them rather than opening and closing the Geoprocessing pane each time a tool is run.

7. **Examine the Favorites tab of the Geoprocessing pane. Commonly used tools are listed at the top and your recent tools below.**

8. **Switch to the Toolboxes tab to see the familiar ArcToolbox and its toolsets. Expand a couple toolsets to see the available tools.**

9. **Switch back to the Favorites tab and select the Buffer tool (figure 6.3).**

10. **Set the tool to buffer the airports using a distance of 50 miles and click Run. Leave the tool open after it finishes.**

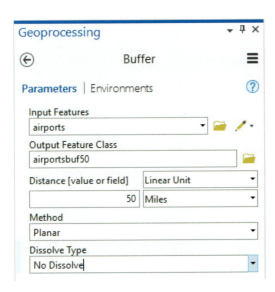

Figure 6.3. Buffering the airports.

TIP Clicking in the Output Feature Class box causes the entire path to appear. Be careful to edit only the last part containing the feature class name or use the Browse button. Click Tab after editing the name to accept the edited name.

11. **On the Analysis tab on the ribbon, in the Geoprocessing group, click the History button to open the History tab of the Catalog pane.**

12. **Point to the top Buffer entry to examine the tool parameters and execution history.**

13. **Return to the Geoprocessing pane, which should still show the Buffer tool with the previous parameters intact.** In many cases, it is best to clear a tool before running it again.

14. **Click the back arrow on the Buffer tool to return to the main Geoprocessing pane.**

15. **Open the Buffer tool again.**

16. **Buffer the highways to a distance of 20 miles, being sure to set Dissolve Type to Dissolve all output features into a single feature.**

17. **Turn off the buffer layers and save the CraterLake project.**

Objective 6.2: Performing interactive selections

Selections in ArcGIS Pro involve similar processes to ArcMap: interactive queries, attribute queries, and spatial queries, although attribute and spatial queries are now accomplished with tools. All three types can be accessed from the Map tab on the ribbon.

1. **Switch to the Map tab and examine the Selection group (figure 6.4a).**

2. **In the Selection group, click the Select drop-down arrow and then the Rectangle tool.**

3. **Select items on the map until you are convinced that this tool is virtually the same as the one in ArcMap.**

4. **Experiment with the other interactive selection tools, such as Circle or Line. Leave some features selected for the next step (don't clear them).**

5. **On the Map tab, in the Selection group, click the Attributes button** 📧 **to open the Attributes pane and see your selected features.**

6. **Clear the selection using the Clear button** 🔽 **in the Selection group.**

7. **To turn off the selection tool, click the Explore tool again.**

    Currently all layers are selectable. ArcGIS Pro users manage what is selectable differently from in ArcMap, by checking layer boxes on the List By Selection tab of the Contents pane.

8. **In the Contents pane, click the List By Selection graphical tab** 🔽 **(figure 6.4b).**

9. **Select some more features using the Rectangle tool.**

10. **Examine the layers in Contents. Find the ones with selected features by looking for the number listed next to the layer name.**

11. **Turn off all the check boxes (use Ctrl+click on one box), and then turn on the check box for counties.**

12. **Draw a box around several counties. Now only counties are selected, and not any other features.**

13. **Turn on selection for another visible layer and try selecting using a rectangle again.**

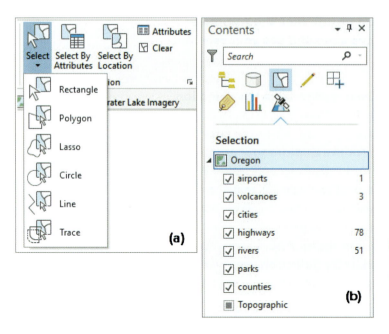

Figure 6.4. Selection tools include (a) interactive selection tools and (b) the List By Selection tab of the Contents pane.

TIP Quickly set a single selectable layer by right-clicking it in the Contents pane and clicking Make this the only selectable layer.

The Selection method and other selection options are now accessed differently as well.

14. **On the Map tab on the ribbon, in the Selection group, click the Options dialog box launcher 🔽 to open the Selection options.**

15. **Examine the options carefully, and then close the window.**

16. **Experiment for a few more minutes until you are comfortable with interactive selection.**

Objective 6.3: Performing selections based on attributes

Selecting features based on an attribute in a table uses a tool now instead of a window, the Select Layer By Attribute tool. The parameters are like the ones in ArcMap, although there is a new way to enter expressions defining the criteria.

1.   **Open the counties attribute table and dock it below the map, if needed.**

2. **On the Map tab on the ribbon, in the Selection group, click the Select By Attributes button to open the Select Layer By Attribute tool (figure 6.5a).**

3. **Set the input layer to counties and click the Add Clause button to enter the expression:** POP2014 is Greater Than 50000, **as shown in figure 6.5a. Click Add, and then click Run.**

4. **Examine the highlighted records in the map and table. Note that the table still includes Show all records** ▤ **and Show selected records** ▤ **option buttons.**

5. **To edit the clause, point to the box containing the clause. A pencil button** 🖊 **appears; click it to edit the clause. Enter 75000 for the new value and click Update. Run the tool again.**

6. **To add a second criterion, click Add Clause again.**

7. **Accept the And operator and enter the expression:** MALES is Less Than FEMALES. **(Click the Fields box above to be able to enter another field to compare with, rather than a value.) Click Add and run the tool.**

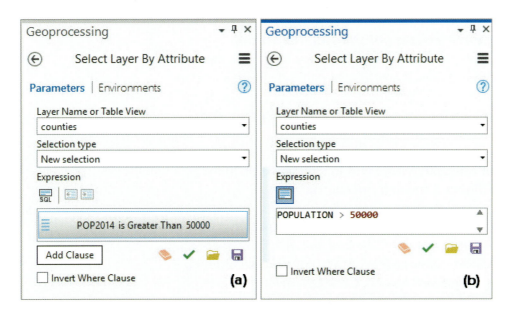

Most people find this new clause builder more intuitive than the ArcMap query builder. If you prefer SQL, however, or encounter a situation where the clause builder does not provide the right expression (such as needing to add parentheses to enforce the

Figure 6.5. Selecting by attributes using (a) the Add Clause feature or (b) an SQL expression.

correct order in a multiclause expression or using SQL functions), you can still edit in SQL (figure 6.5b).

8. **In the Select Layer By Attribute tool, click the SQL button** **in the Expression section.**

9. **Edit and execute a new expression, if you want.**

10. **Switch back to the clause builder by clicking the SQL button again.**

11. **Clear the selection using the Clear button** **on the Map tab on the ribbon, in the Selection group, or the Clear Selection button** **in the table.**

12. **Close the counties table.**

TIP *Clear a selection by mistake? No problem. If the selection tool is still open, simply run it again with the same parameters. If it's no longer open, return to the main Geoprocessing pane and use the Recent section to reopen and rerun the tool.*

Objective 6.4: Performing selections based on location

Performing spatial queries is now also accomplished with a tool.

1. **On the Map tab on the ribbon, in the Selection group, click Select By Location.**

2. **Set the tool to select highways within a distance of** 30 miles **of volcanoes (figure 6.6a) and run the tool.**

TIP *ArcGIS Pro versions prior to 2.1.1 did not have the ability to create a new layer from a set of selected features. This absence has been remedied, although it might still have bugs. If nothing happens, save the project, reopen it, and try again.*

3. **Save the CraterLake project, just in case.**

4. **In the Contents pane, right-click the highways layer and click Selection > Make Layer From Selected Features. When the new layer appears, rename it** Hazardous Highways.

TIP The Make Feature Layer tool can be used to select by attributes and create a new layer in a single step (figure 6.6b).

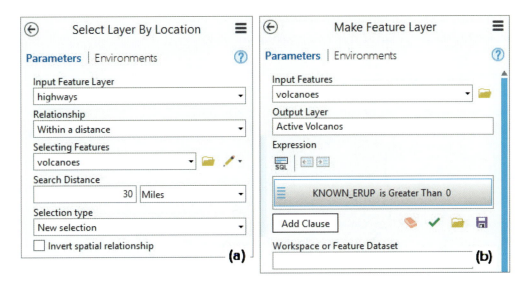

Figure 6.6. Creating a new layer by (a) selecting highways near volcanoes and (b) making a layer from active volcanoes.

Objective 6.5: Practicing geoprocessing

In this section, we will run through a short analysis to practice the geoprocessing techniques just outlined. The goal of the project includes finding which sections of interstates and which cities are at risk from active volcanoes in Oregon.

1. **Click the back arrow to return to the main Geoprocessing pane.**

2. **Start typing** make feature **in the Search box to search for the Make Feature Layer tool. Click it to open it.**

3. **Use the tool to create a layer of active volcanoes, those with KNOWN_ERUP greater than zero (figure 6.6b).**

4. **Buffer the active volcanoes to a distance of** 60 miles.

5. **Select only the interstates from the highways layer.**

6. **Clip the interstates with the volcano buffers to isolate the interstates near volcanoes.**

7. **Select the cities within 60 miles of the active volcanoes and make a layer from them.**

8. **Symbolize the map to show the at-risk cities and interstates (figure 6.7).** The analysis shows there are about 187 cities in the hazard zone.

9. **Save the CraterLake project.**

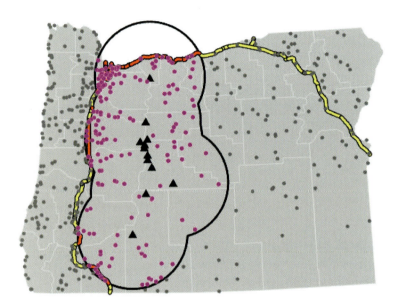

Figure 6.7. Results of the volcanic hazards analysis.

Chapter 7

Tables

Background

Tables in ArcGIS Pro, like feature classes and rasters, use the same data model and can be managed by either ArcMap or ArcGIS Pro. However, the way that they are managed has changed in many ways.

General table characteristics

Like maps and scenes, tables in ArcGIS Pro are displayed as views in the central display area of the program. As views, they can be docked, stacked, or displayed side by side with each other or with other types of views. The same docking and placement tools used for other views pertain to tables as well.

Attribute tables associated with spatial datasets in a map can be opened from the Contents pane. Standalone tables, if present in the map, appear at the bottom of the Contents pane. Unlike ArcMap, standalone tables appear at the bottom of the List By Drawing Order tab, with no need to open the List By Source tab to see them. Also like layers, a table view is a slightly different entity than the underlying source table. The relationship of a layer to its source data is analogous to the relationship between a table view and its source table, in that the table has properties that can be set without modifying the source table (such as sorting the records).

Some of the familiar table functions, available as menu options or as a right-click, are missing or significantly changed (figure 7.1). The Statistics menu item that appears when a field is right-clicked, for example, now opens a chart that can be manipulated and saved, instead of a temporary window. More advanced statistics can be calculated by running the Summary Statistics tool and producing a permanent output table. Functions in ArcMap that operated on tables, such as selecting, joining, or calculating, now open geoprocessing tools rather than menus.

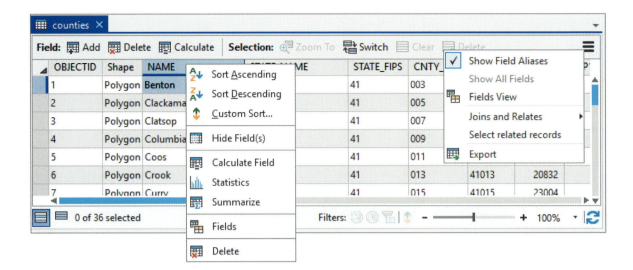

Additional settings for tables are accessed from two different tabs on the ribbon: the Data tab of the Feature Layer contextual tab (figure 7.2a) and the Table: View tab (figure 7.2b). Some of the items are the same on both tabs and in the table view, or perform the same function.

Figure 7.1. A table view showing the open field heading menu and the Options menu.

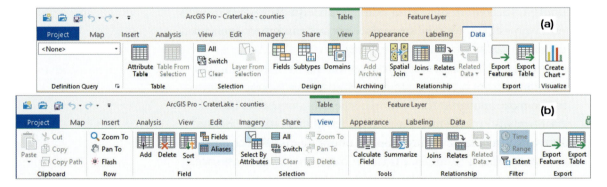

Figure 7.2. Ribbon tab functions for tables.

Joining and relating tables

The joins in ArcGIS Pro, like many other previous menu items in ArcMap, have been implemented as geoprocessing tools. Even though the menu item may be visible on the Contents pane or in the table, selecting it opens the tool (figure 7.3a). Except for this difference, attribute joins are virtually identical. However, the terminology has changed. The table that receives the data, previously termed the *destination table*, is now called the *target table*. The table providing the additional information is now known as the *join table* instead of the *source table*.

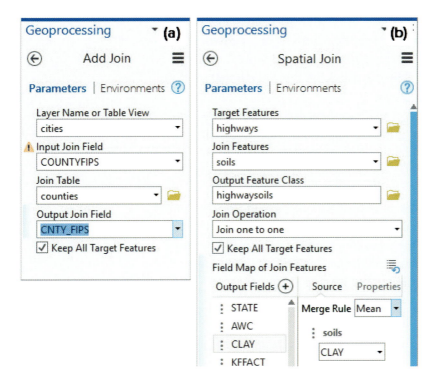

Figure 7.3. ArcGIS Pro join tools include (a) the Add Join tool and (b) the Spatial Join tool.

Spatial joins have changed even more. They have also been implemented as a tool. The ArcMap spatial join menu was a relatively simple affair, allowing two spatial conditions (inside or distance) and two cardinality options (simple or summarized), for a total of four basic options, and even that level of complexity was not easy to understand. The Spatial Join tool offered more options but was confusing at first.

In ArcGIS Pro, the simpler menu is gone, and only the Spatial Join tool is available (figure 7.3b). It offers the full spectrum of spatial operators (intersect, within, contains, within a distance, closest, and so on). It also offers two join types: one to many and one to one. The terminology used is particularly confusing because similar terms are used to describe cardinality between tables, with different meanings. A brief explanation follows, but if you are unfamiliar with spatial joins or cardinality, you might want to simply skip to the next section for now.

When joining two tables, the cardinality between them is paramount and dictates the structure of the join. If one record in the target table (destination) matches one record in the join table (source), it is termed a one-to-one cardinality. If one record in the target table matches many records in the join table (for example, states to cities), it is one-to-many cardinality. If many records in the target table match one record in the join table (cities to states), it is many-to-one cardinality. Table joins can be executed with only a one-to-one or many-to-one cardinality; a one-to-many cardinality requires a relate instead. In ArcMap, a one-to-many cardinality requires a relate that associates the tables but does not actually combine them into one structure.

In the Spatial Join tool, there are two join types, called *one to one* and *one to many*, but it means something quite different. The join type controls what happens when a single record in the target table matches many records in the join table (a one-to-many cardinality).

In a spatial join type of one to many, duplicate output features are created until there are enough of them to accommodate each matching join feature, so that every join record can be matched to a target record. The output feature class may then contain many more features than the original target feature class.

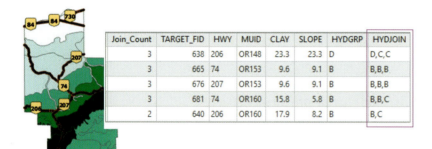

Figure 7.4. Joining soil information to highways using a spatial join of one to one.

TIP *It is important to distinguish the join type from the cardinality between the tables, which are two different concepts but with similar terms. A join of one to one is frequently performed on tables with a one-to-many cardinality.*

In a spatial join type of one to one, one output feature is created for each target input feature. If the target feature matches more than one feature in the join table, the matching records are summarized using merge rules to create a single record to append to the target feature (for example, by averaging or summing numeric values). The output feature class will contain the same number of features as the original target table (or fewer, if features without matches are dropped).

TIP *The join type matters only when the cardinality between the tables is one-to-many cardinality; in other cases, the two join types give the same result.*

Consider the situation of a spatial join between highways (the target features) and soils in figure 7.4, in which a one-to-many cardinality exists between each highway segment and the soil polygons it crosses. Figure 7.3b shows the tool configuration that produced the result in figure 7.4. A join type of one to one was selected so that the output feature class contains the same number of highway features as the input target table. In the output table, the Join_Count field indicates how many soil polygons each highway crossed. The CLAY and SLOPE fields were assigned a merge rule that calculated the average value of the units crossed (the merge rule for CLAY is shown in figure 7.3b). The text fields MUID and HYDGRP were assigned the first value crossed. A new HYDJOIN field was created with a concatenate (join) merge rule, so that the HYDGRP value of each soil unit was appended to the field, providing a list of values from the units crossed.

Making charts

Creating charts in ArcGIS Pro is different from creating them in ArcMap, although the same chart types are available, and the live link feature between the chart, table, and map, which is such a benefit for data exploration, remains intact. The process is simpler and has some new perks, such as additional chart types and statistics.

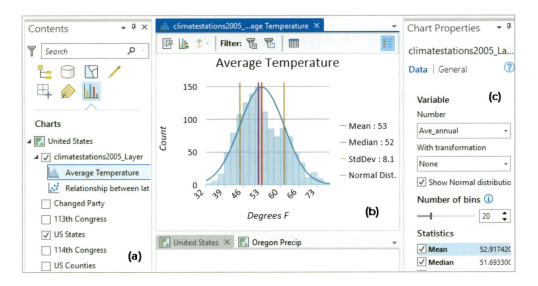

Figure 7.5. Making charts in ArcGIS Pro utilizes (a) the Charts tab of the Contents pane, (b) a chart, and (c) the Chart Properties pane.

Charts are viewed and managed using the Charts tab of the Contents pane (figure 7.5a). All charts associated with a specific map are shown in this pane. This United States map has two charts that have been created from a climatestations layer. One, the Average Temperature histogram, is open and is visible in figure 7.5b. It includes the mean, median, and standard deviation, as well as the curve representing a normal distribution. Figure 7.5c shows the Chart Properties pane, which is used to set the properties for a chart, such as the field being graphed and the number of bins.

Time to explore

For the next few exercises, we will use the CraterLake project and manipulate the tables in the Oregon feature dataset. You will become familiar with table views and how they are organized, learn to make charts, and explore the new ways to generate statistics.

Objective 7.1: Managing table views

Managing tables in ArcGIS Pro is not that different from using them in ArcMap.

1. **Open the CraterLake project.**

2. **Close any open maps except for the Oregon map. Open the List By Drawing Order tab 🗂 in the Contents pane.**

3. **Open the table for the counties layer.**

4. **Explore the menus and buttons in the table view (see figure 7.1).**

5. **Try sorting the table by right-clicking a field.**

   Several tabs on the ribbon contain functions for working with tables.

6. **In the Contents pane, make sure the counties layer is highlighted. Explore the buttons on the Feature Layer: Data tab on the ribbon.**

7. **Click the counties table to select it and examine the Table: View tab on the ribbon.**

8. **On the Table: View tab, in the Field group, click Fields.** In ArcGIS Pro, the Fields properties menu has been replaced by the Fields view. It can be used to make a variety of cosmetic changes to a table—or more precisely, to the table view. You can turn off certain fields to focus attention on the remaining ones, or change the number format of a field's values to make them easier to read—for example, by adding a comma, such as 198,000 instead of 198376.

9. **In the Fields view, use the Visible check box at the top to clear the check boxes for every field at once. Then click the individual check boxes for STATE_NAME and POPULATION to turn on only those two fields.**

10. **Under Visible, in the row for the POPULATION field, click in the Number Format box (it currently says Numeric) to make the ellipsis (…) appear. Click it to open the Number Format options.**

11. **Set the options to 3 significant digits and show thousands separators. Click OK.**

12. **Because the Fields view is open, the Fields tab is visible on the ribbon. Open it to click the Save button in the Changes group. Then close the Fields view.**

13. **Examine the changes in the table (close and reopen it, if needed, to see them).**

TIP *Make sure to save changes before leaving the Fields view, or you may find you are unable to perform other operations. Also, it is wise to close the Fields view when not working with it.*

14. **Reopen the Fields view for counties and make all the fields visible again.**

15. **Click on the left of the field name and drag POPULATION above STATE_FIPS to change the field order.**

16. **Save the changes and close the Fields view.**

17. **Open the table for the parks layer.** Table views are arranged and docked the same way that other views are managed.

18. **Place the parks table side by side with the counties table (figure 7.6).**

19. **Practice arranging the two tables in various configurations with the map and with each other.**

20. **Close all the tables when finished.**

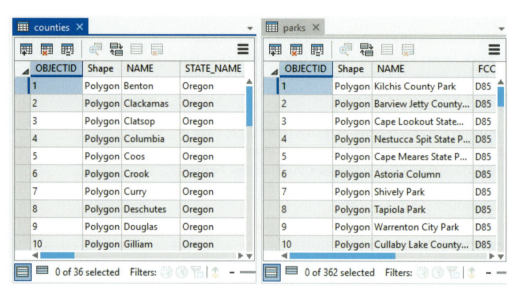

Figure 7.6. Docking skills are used to arrange tables side by side.

Objective 7.2: Creating and managing properties of a chart

Now you can create a histogram for the elevation of Oregon's cities.

TIP The chart will be created for the layer selected in the Contents pane.

1. **In the Contents pane, select the cities layer.**

2. **On the ribbon, on the Feature Layer: Data tab, in the Visualize group, click Create Chart > Histogram.** The Chart Properties pane opens automatically (figure 7.7).

3. **Choose ELEV_IN_FT for the Number field.**

4. **Experiment with the settings in the pane.**

5. **Switch to the General tab and give the chart a better title and x-axis label.**

6. **Close the chart view.**

7. **In the Contents pane, switch to the List By Charts tab** **and find the elevation histogram.** The List By Charts tab of the Contents pane provides the means to view, organize, and create charts.

8. **Right-click the counties layer and click Create Chart > Scatter Plot.**

9. **In the Chart Properties pane, choose MED_AGE_M for X-axis Number and MED_AGE_F for Y-axis Number.**

10. **Select a group of points on the lower end of the trend line and view the highlighted counties on the map. Which regions in Oregon have younger populations? Which have older populations?**

11. **On the ribbon, on the Map tab, in the Selection group, use the Select tool to select features on the map and see them highlighted in the plot.**

12. **Examine the statistics and experiment with the pane settings.**

13. **Try different combinations of variables to plot.**

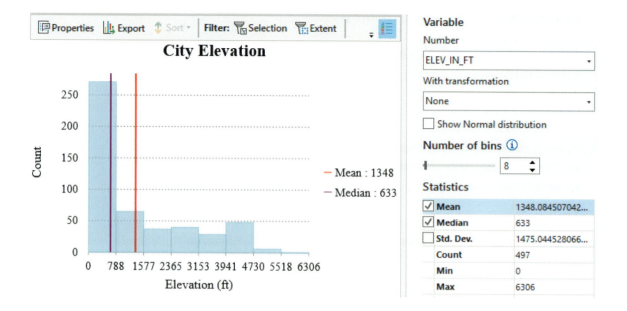

Figure 7.7. Creating a histogram for city elevations.

These charts are meant more as exploratory tools than publishing tools, but some chart settings can be modified.

14. **Make sure that a chart is the active view (by clicking its title bar) and open the Chart: Format tab on the ribbon.** The Current Selection group on the ribbon controls which element of the chart is being modified. Use the Selection drop-down arrow to choose All Text (scroll up if needed).

15. **Change the font of the text.**

16. **Experiment with the other settings on the Chart: Format tab.**

17. **Experiment with a few more chart types if you have time.**

18. **Close all the charts and save the CraterLake project.**

Objective 7.3: Calculating statistics for tables

The familiar Statistics menu item in a table now creates a chart with a mean.

1. **Open the attribute table for counties.**

2. **Right-click the POP2014 field heading and view the options. Then choose Statistics.** A new histogram chart is created.

3. **Examine the statistics in the Chart Properties pane, and then close the chart view.**

4. **On the ribbon, use the Map tab, in the Selection group, to clear any selections you may have.**

5. **In the Geoprocessing pane, search for and open the Summary Statistics tool.** The Summary Statistics tool can be used to calculate statistics for one or more fields in a table, with or without a case field to group the table records first.

6. **Set the input table to counties and the output table to CountyMinorities.**

7. **Click the Field drop-down arrow and check the boxes for these minorities: AMERI_ES, ASIAN, BLACK, HAWN_PI, HISPANIC, MULT_RACE, and WHITE (figure 7.8a). Click Add.**

8. **Accept the Sum statistic for each minority.**

9. **Click Run.**

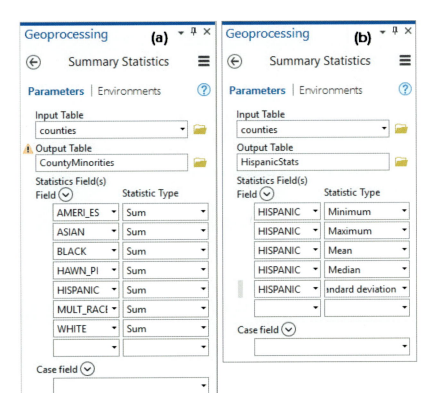

Figure 7.8. Calculating statistics for fields in a table using the Summary Statistics tool: (a) sums for each minority and (b) different statistics for Hispanics.

The output table appears at the bottom of the Contents pane under Standalone Tables.

10. Open and examine it.

11. Close the table.

12. Clear and reopen the Summary Statistics tool. This time, set it up to calculate the minimum, maximum, mean, and standard deviation of just the Hispanic minority (figure 7.8b). Run it.

The Summarize function in ArcMap has been replaced by this Summary Statistics tool, run with a case field to group the features. Let's determine the statistics about city dwellers for each county, grouping the cities by county before calculating the statistics.

13. Clear and reopen the Summary Statistics tool.

14. Set the input table to cities and the output table to CityStatsByCounty.

15. Set the statistics as in figure 7.8b but using the cities POP2010 field and adding the Count and Range statistics.

16. Set the Case field to COUNTY. Run the tool.

17. Open the output table. Evidently, many of the cities are missing population values, so the Minimum and Range fields are meaningless.

18. Right-click the MIN_POP_2010 field and click Delete. When asked whether to delete, click Yes.

Objective 7.4: Calculating and editing in tables

Unlike ArcMap, layers and tables in ArcGIS Pro are always editable by default, and no edit session need be opened. You'll learn more about this feature in "Editing," chapter 10, but let's use it now to make a quick edit to a table. The Range statistic in this table is flawed by the –999 flags, so it's a good place to practice editing without fear of damaging valuable data.

1. **Click a cell in the RANGE_POP_2010 field of the CityStatsByCounty table and change its value.**

2. **Change a few more cell values for practice.**

3. **To save the changes, switch to the Edit tab on the ribbon. In the Manage Edits group, click Save to save the edits or click Discard to discard them.**

As with many other functions in ArcGIS Pro, calculating values into a field requires a tool. Let's calculate a slightly better (but still flawed) RANGE value for the cities by assuming that the minimum population is zero instead of the flag value –999.

4. **Right-click the RANGE_POP_2010 field heading and click Calculate Field. The Calculate Field geoprocessing tool opens (figure 7.9).**

5. **Double-click items in the Fields list and the operators list to enter this expression: !MAX_POP_2010! - 0.**

6. **Check your expression against the expression in figure 7.9.**

TIP The tool automatically adds exclamation points (!) around each field name if you double-click the name from the list. If typing the expression, you must enter them yourself.

7. **Examine the rest of the tool, noting its similarities to the function in ArcMap, and its differences.**

8. **Run the tool and examine the results.**

9. Save the CraterLake project.

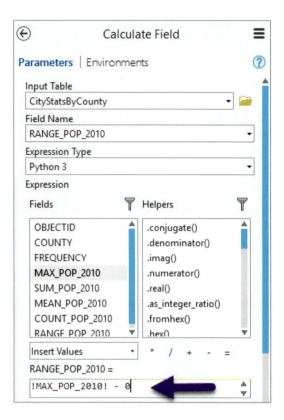

Figure 7.9. Calculating a field.

Chapter 8

Layouts

Background

ArcGIS Pro was designed to achieve a long-desired objective for many ArcMap users: the ability to create multiple layouts from one map, rather than being restricted to a single layout per map document.

Layouts and map frames

A project may contain as many layouts as desired, each one built from one or more maps residing within the project. Recall that an ArcGIS Pro map is the functional equivalent of the ArcMap data frame. However, when you can create multiple layouts using the same maps, a problem arises with specifying the map scale and extent. If these settings are shared by the map and the layout, zooming in or out of the map would disrupt the layout, or multiple layouts. In ArcGIS Pro, a method was thus devised to allow the layout map extent to be independent of the map, and of other layouts.

This method uses a *map frame* to accomplish this objective. When a map is placed in a layout, it becomes a map frame, which must be activated to set the scale and extent. Once they are set and saved, and the map frame is deactivated, the scale and extent remain fixed within the layout. In this way, the same map can also have multiple extents in different layouts.

Within a data frame, ArcMap offered three ways to control the behavior of a map inside the data frame when it was resized on the page: automatic, fixed scale, and fixed extent. ArcGIS Pro has a similar function but includes five different methods:

1. **None** indicates no scaling and is the equivalent of the automatic setting in ArcMap.
2. **Fixed extent** sets the exact extent (or position of the camera for a 3D scene). The map frame may be resized but must conform to the same aspect ratio. All navigation is disabled. The

extent may be set to the current extent of the frame, from one or more of the layers in the frame, or from a rectangle derived from specified coordinates.

3. **Fixed center** fixes the center point of the view, permitting zooming and rotating (3D) around the center of the map extent. The panning function is disabled.

4. **Fixed center and scale** fixes the center point of the view as well as the scale of the map frame. Zoom and pan are both disabled, but rotation is allowed. This option is similar to fixed extent but allows the map frame itself to be resized to a different aspect ratio.

5. **Fixed scale** fixes the map scale to a specific value, such as 1:24,000. The frame may be resized, which will increase the geographic area of the map displayed. Zooming is disabled, but panning is permitted.

The additional map elements, or *map surrounds*, offered by ArcGIS Pro correspond to those in ArcMap: text, legend, scale bar, north arrow, graticules, and so on. However, the editing interface is quite different and confusing at first, but it does have a logic that sinks in after some practice. ArcMap users may initially be frustrated but, over time, come to appreciate the new system.

Layout editing procedures

To become acquainted with the layout functions in ArcGIS Pro, look at a completed layout as shown in figure 8.1, and we'll visit the GUI elements one by one. The layout occupies the center of the GUI in a layout view. This layout contains two map frames, a Hazards map frame showing Washington and Oregon and a USA map frame used as a location map. Each map frame, in turn, is based on a map in the project.

First look at the Contents pane ([1] in figure 8.1). When a layout is open, the Contents pane reflects the contents of the layout, including all map frames and map surrounds. Notice entries for the legend, a scale bar, a north arrow, a title, and text. When an item is selected in the Contents pane (as the scale bar is currently selected), its properties can be opened and manipulated in a Format pane (2), shown on the right of the screen. Items can also be selected by double-clicking them on the layout, or right-clicking them to open their properties (although this can be difficult when elements lie on top of each other, so the Contents pane is often easier to use). Notice the handles around the scale bar in the map frame (3), indicating that it is currently selected.

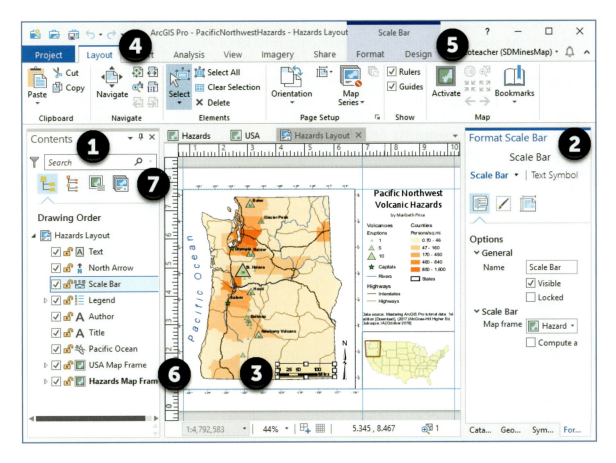

Above the layout view, the Layout tab (4) is currently open on the ribbon, showing tools for manipulating the layout. Notice also the Scale Bar contextual tab set containing two more tabs, Format and Design (5). Many elements in the Contents pane will, when selected, cause contextual tabs to appear. The most common and basic settings are managed on the ribbon, and more advanced settings are typically available in the Format pane.

The Contents pane of a layout also includes the maps upon which the layout is based (6). The two map frames, Hazards Map Frame and USA Map Frame, can be expanded to show the maps and manipulate the settings normally associated with the map layers, such as symbols, labels, and so on.

Figure 8.1. The components of the layout GUI. See the text for the explanation of the items indicated by each circled number.

TIP *Although the map frame scale and extent in a layout are independent of the map, the symbols and labels are not. Changes made to symbols, labels, and other layout properties will be reflected in all layouts based on that map, and in the map itself.*

Notice that the Contents pane has multiple graphical tabs that present the pane in different ways (7), including, from left: List By Drawing Order, List By Element Type, List By Map Frame, and List Map Series Pages.

Importing map documents and templates

ArcGIS Pro includes the ability to import a map document, so favorite maps and templates from ArcMap can be easily brought into a project. In fact, the Hazards map of Washington and Oregon shown in figure 8.1 was initially imported from a map document. All aspects of the map document layout will be integrated, including folder connections, page settings, extents, and even styles and symbols. The styles are particularly helpful, because ArcGIS Pro uses a different method of storing styles, and the additional style sets (Civic, Geology 24k, Forestry, and so on) that were available in ArcMap no longer appear in ArcGIS Pro, although they may be imported and used.

The Insert tab on the ribbon has a button in the Project group for importing a map document. It will appear as a new layout in the Layouts folder of the Catalog pane. Once imported, however, the map layout cannot be exported back to a map document for use in ArcMap. Any changes to the layout in ArcGIS Pro must remain in ArcGIS Pro.

Map templates were a commonly used feature in ArcMap that have no precise equivalent in ArcGIS Pro. However, a layout may be saved as a layout file (.pagx) to provide a functional equivalent. It includes the page setup, the elements of the layout such as legends and graticules, and any maps referenced by the map frames. However, it does not include any data referenced by the maps, so if the data has become unavailable, the maps will be blank and the data sources must be reestablished. A layout file may be shared with other ArcGIS Pro users, but it cannot be opened in ArcMap or, depending on the version number, in some versions of ArcGIS Pro previous to the one in which it was created.

Time to explore

You'll use the CraterLake geodatabase to learn to create and edit layouts. The overall order of steps to create a layout may be summarized as follows:

- A new layout is created and named. It will be saved with the project.
- The map page size is configured, including rulers and guides, if desired.
- The map frames are chosen and added to the layout.
- The map frames are activated to set the scale and extent.
- Map surrounds are added using the Insert tab.
- Properties of map frames and surrounds are manipulated using tabs on the ribbon and settings in the Format pane to achieve the desired effects.

Objective 8.1: Creating the maps for the layout

The first step of any layout includes configuring the maps that it will contain. The main map for this layout will be based on the Crater Lake map of the CraterLake project.

1. **Open the CraterLake project.**

2. **Open the Crater Lake map and close all other maps.**

3. **Turn on the Hillshade layer. Select the Crater Lake Geology group layer and use the Layer: Appearance tab on the ribbon, and the Effects group, to make it about 25% transparent.**

4. **Adjust any other settings, if needed, to make a nice map.**

   Now you can make a location map as well.

5. **On the ribbon, use the Insert tab, and the Project group, to create a new map. Rename it Location.**

6. **Add the counties feature class from the Oregon feature dataset in the CraterLake geodatabase.**

7. **Symbolize the counties with a hollow symbol and dark-gray outline.**

8. **Turn off the basemap.**

9. Save the CraterLake project and close the maps.

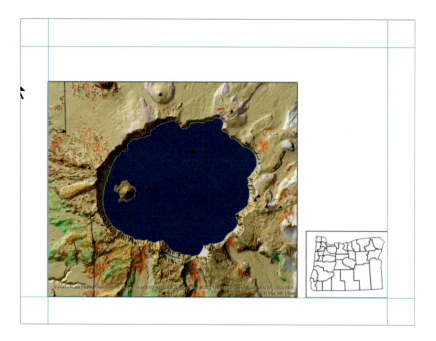

Figure 8.2. Arranging the map frames in the layout.

Objective 8.2: Setting up a layout page with map frames

Next, you'll create the layout and assign the maps to it.

1. **On the ribbon, open the Insert tab and, in the Project group, click the New Layout drop-down arrow. Select the ANSI – Landscape Letter template.**

2. **Rename the layout** Visitor Map**.**

3. **Open the Insert tab. Notice that it now contains many buttons for inserting map frames and map surrounds, such as north arrows or scale bars.**

4. **In the Map Frames group, click the Map Frame drop-down arrow and choose the Crater Lake map. Depending on the version of ArcGIS Pro, the frame may appear automatically, or you may need to draw a box for the frame.**

5. **While it is still selected on the layout, use the selection handles to resize it to cover the left two-thirds of the page, leaving healthy margins.**

6. **Click the Map Frame drop-down arrow again, scroll down, and choose the Location map (Default).**

7. **Arrange the maps approximately as shown in figure 8.2.**

8. **In the Contents pane, click twice on the lower map frame and note that the Crater Lake map is highlighted. Rename the Map Frame to Crater Lake Fr.** The maps were loaded into the Contents pane with generic names. Editing is easier if you give better names to the layout elements.

9. **Rename the other map frame Location Fr.**

10. **Expand the Crater Lake Fr map frame entry, if needed.**

TIP Be sure to differentiate between the map frame (a map placed on the layout) and the map. This distinction is important regarding the steps in the objectives that follow.

The entire map, with all its layers, is included in the Contents pane. You could use these entries to modify symbols or labels, if needed, without opening the original maps.

11. **Collapse the Crater Lake Fr map frame for now.**

Objective 8.3: Setting map frame extent and scale

The map frame extent and scale are controlled by the size of the map frame on the page, the extent of the map inside the map frame, and the scale assigned to the map.

1. **With the Crater Lake Fr map frame still selected in Contents, open the Map Frame: Format tab on the ribbon.**

2. **Examine the drop-down box in the Current Selection group. With it set to Map Frame, only the Size & Position group is active.**

3. **Set X and Y to 0.75 in, Width to 7 in, and Height to 6 in.**

Guides are helpful in aligning elements along margins or other lines. You can add one guide at a time by right-clicking a location on the ruler and choosing Add Guide, but here you'll add a complete set of margin guides at the same time.

4. **On the map display, right-click the ruler and click Add Guides.**

5. **Choose Both for Orientation, Offset from edge for Placement, and enter** 0.75 in **as the Margin distance. Click OK.**

6. **Adjust the Location Fr map frame to fit within the guides.**

7. **In Contents, select the Crater Lake Fr map frame.** Now that the frames are in place, it is time to set the extents inside them.

8. **On the ribbon, open the Layout tab and choose Activate.**

9. **On the ribbon, on the Map tab, in the Navigate group, use the Explore tool to zoom/pan inside the frame until the extent looks good. (Recall that right-clicking and dragging the mouse is used for continuous zooming.)**

10. **Return to the Activated Map Frame: Layout tab on the ribbon and, in the Map group, click Close Activation.**

11. **Repeat the procedure to set the extent for the Location Fr map frame, centering the Oregon counties in its frame.** The layout should look similar to figure 8.2.

Objective 8.4: Formatting the map frame

Recall that the tabs on the ribbon generally present common settings but the Format pane has more detailed ones.

1. **In Contents, select the Crater Lake Fr map frame.**

2. **On the ribbon, open the Map Frame: Format tab.**

3. **In the Current Selection group, change the drop-down arrow from Map Frame to Border.**

4. **Change the width or color of the map frame border.**

5. **In Contents, select the Location Fr map frame. Change its border to match.**

   Next, take a tour of the more detailed properties in the Element pane. It will take some practice to learn where everything is. The title presented at the top of the Element page changes, depending on what type of element you are formatting. It is helpful to keep the Element pane open throughout the layout editing process, and convenient to dock it on top of the Catalog pane.

6. **In Contents, right-click the Crater Lake Fr map frame and click Properties to open the Format Map Frame mode of the Element pane.**

7. **Select one of the map frame characteristics from the main drop-down menu: Map Frame, Background, Border, or Shadow.**

8. **Examine the graphical secondary tabs of the pane. Use the icons to select every tab and review the settings each one contains.**

9. **Change the drop-down menu to another characteristic. Continue until you have reviewed all the settings for each one.**

TIP Many actions open the Element pane. Right-click an element in the Contents pane and click Properties, or double-click it as a short cut. Select and right-click an element in the map, or double-click it. Once the Element pane is open, click any element to switch the pane to format it.

   Placing an extent indicator in a map is simple.

10. **In Contents or on the map, select the Location Fr map frame.**

11. **Open the Insert tab on the ribbon and, in the Map Frame group, click the Extent Indicator drop-down arrow. Select the Crater Lake Fr map frame.** The extent box is tiny, but it's there.

12. **To show a point instead, expand the Collapse to point heading (figure 8.3) in the Element: Format Extent Indicator mode that automatically opened. Underneath, increase the Smaller than setting to about** 5 pt.

13. **Save the CraterLake project.**

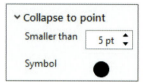

Figure 8.3. Collapsing an extent rectangle to a point.

Objective 8.5: Creating and formatting map elements

Now that you have the basic map frames set up, it's time to add some surrounds. You'll start with a title.

1. **On the ribbon, open the Insert tab and, in the Text group, click the Text button A.**

2. **Click above the map to place the text. While the text is still highlighted, type** Crater Lake Visitor Map, **and then click somewhere outside the text to finish and select it.**

3. **Open the Format Text mode of the Element pane and set the font and size.**

4. **On the map display, drag the text to place it as desired.**

5. **In Contents, rename the Text entry** Title **to make it easier to recognize.**

TIP You can also drag the selection handles of a text box to change the size of the text.

Next, add a north arrow to the layout.

6. **On the ribbon, open the Insert tab and, in the Map Surrounds group, click the North Arrow drop-down arrow and choose a symbol.** It is automatically placed near the center of the layout.

7. **Drag the north arrow to the desired location.**

8. **To modify the north arrow, look on the ribbon for the North Arrow contextual tab. Open the North Arrow: Format tab and examine its contents. Also view the contents of the North Arrow: Design tab.**

9. **Use the tabs to modify the north arrow properties, if desired.**

   Next, you will add a scale bar. Be careful to create it for the correct map frame containing Crater Lake.

10. **In the Contents pane, select the Crater Lake Fr map frame.**

11. **On the ribbon, open the Insert tab. In the Map Surrounds group, click the Scale Bar drop-down arrow to choose a style. It will automatically be placed on the map. Drag it to the desired location.**

12. **On the ribbon, examine the Scale Bar: Format and Scale Bar: Design tabs. Most of these settings will be like those found in ArcMap. Experiment with them to modify the scale.** On the Scale Bar: Format tab, don't forget to examine the Current Selection drop-down menu to see what else can be modified.

   Finally, the map needs a legend. Again, be careful to select the correct map frame before creating the legend.

13. **In the Contents pane, select the Crater Lake Fr map frame.**

14. **On the ribbon, open the Insert tab and, in the Map Surrounds group, click the Legend button.** The shape of the mouse pointer changes.

15. **Drag a rectangle in the empty space on the right of the main map and above the location map, as shown in figure 8.4. When you release the mouse button, the legend will be drawn.**

16. **Rotate the wheel button on the mouse to zoom in for a better view of the legend. You can also click the wheel button and drag to pan to the legend.**

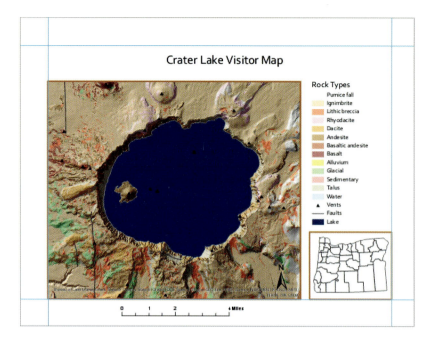

Crater Lake Visitor Map

Figure 8.4. The final layout.

Objective 8.6: Fine-tuning the legend

You can make a few changes to perfect the legend. The rock units are portrayed twice, once for the crater floor and again for the area around the lake. You can turn the crater floor units off in the map itself; they can't be seen anyway because the lake covers them. You can also remove the All Other Values label because there are no items in it.

1. **In Contents, expand the Crater Lake Fr map frame and the Crater Lake map, finding the group layer Crater Lake Geology.**

2. **Click to clear the check box to turn off the Crater Floor Geology layer.**

3. **Click the Rock Types layer to select it and open the Symbology pane.**

4. **In the Symbology pane, click the More drop-down arrow and click Show all other values to turn it off.**

We want to keep displaying the Hillshade layer, but you can turn it off in the legend.

5. **In Contents, expand the Legend entry. Click to clear the Hillshade check box.**

6. **Click the Rock Types layer underneath the Legend entry and drag it to the top of the other items so it is shown first (figure 8.5).** The order of layers listed in the Contents pane dictates the order of layers in the legend.

Figure 8.5. Changing the legend order by dragging the layers.

You don't need so many headings above the Rock Types layer. You can remove them in the Format Legend mode of the Element pane so it will appear as in figure 8.5.

7. **In Contents, right-click the Rock Types entry under the Legend heading and click Properties.**

8. **In the Format Legend Item mode of the Element pane, under the Show heading, clear all the check boxes except Layer Name and Label.**

9. **Click somewhere off the legend to deselect it, or click something else.**

10. **Zoom back to the full extent of the layout using the mouse wheel (or open the Layout tab and, in the Navigate group, use the Explore tool). The layout should look similar to figure 8.4.** Notice how simple it is to work with the layout elements. When in doubt, find the item to change in Contents, select it, and use the contextual tabs, or right-click it and use properties to show the Element pane. The Element pane will often open to the settings you need, but if not, you can explore the pane until you find them.

11. Save the CraterLake project.

12. Close the Layout view.

Objective 8.7: Accessing and copying layouts

Layouts can be accessed from a folder in the Catalog pane, just like maps.

1. **Switch to the Catalog pane.**

2. **Expand the Layouts entry to see the layout just created.**

3. **Double-click the Visitor Map layout to open it again.**

   Do you need to create a new layout similar to one you already have (as you did by using Save As on map documents over and over)? Nothing could be simpler.

4. **In the Catalog pane, right-click the Visitor Map layout and click Copy.**

5. **Right-click the Layouts folder and click Paste.**

6. **Right-click the new Visitor Map1 entry and click Rename to call it something else.**

7. **Save the CraterLake project.** We have now covered the basic skills needed to create and edit layouts. Now all you need is practice. Play around with the interface, experiment with the settings, and make messes with practice maps until creating a good layout starts to become second nature.

Chapter 9

Managing data

Background

ArcGIS Pro relies on the same data models that ArcMap does, including geodatabases and shapefiles, as well as the same raster formats. Although project documents are not backward-compatible, the datasets created using ArcGIS Pro are compatible with ArcMap.

Data models

Since you might have to alternate between ArcMap and ArcGIS Pro for a time, you will be glad to know that you can access the same datasets within a geodatabase from either program, edit them with either program, edit them in one program and then with another (but never at the same time), and so on. You can even update the structure, or schema, in ArcGIS Pro using tools not available in ArcMap, and ArcMap has no trouble reading the result. For example, ArcGIS Pro allows you to rename a field in a table. ArcMap cannot, but it can read the feature classes once ArcGIS Pro has finished changing them. (However, map documents based on those datasets may no longer function correctly, so use caution.)

One major difference for ArcGIS Pro, however, is that it neither recognizes nor works with the older data models, including ArcInfo coverages and personal geodatabases. Anyone storing information in those formats will be required to upgrade them to a file geodatabase or enterprise geodatabase format for use in ArcGIS Pro, and this conversion must be done using ArcMap or ArcCatalog.

ArcGIS Pro has caught up with ArcCatalog for metadata management as of version 2.2. You can edit metadata using the ArcGIS Pro Catalog view, creating it using one of the standard metadata styles such as Federal Geographic Data Committee Content

Standard for Digital Geospatial Metadata (FGDC-CSDGM) or International Organization for Standardization ISO 19115. The metadata can also be exported to an XML format file to create compliant standards-based metadata. Starting with ArcGIS Pro 2.2, the software can read and update existing CSDGM metadata for a dataset, converting it to the internal ArcGIS format so that it can be viewed by ArcGIS Pro.

Another precept to remember is that the Catalog pane and Catalog view in ArcGIS Pro are, by design, somewhat less flexible than ArcCatalog. For example, many tasks associated with a right-click in ArcCatalog may have been relegated to tools and cannot be accomplished any other way, although some of the early omissions, such as creating a folder, are making their way into later versions of the software. If having trouble finding a function that exists in ArcCatalog, try searching for a tool version in the Geoprocessing pane. If it still cannot be found, it may also not yet be available in the current version of ArcGIS Pro. ArcGIS Pro Help won't inform you when to expect this functionality to arrive, but it may help confirm that the function isn't available yet and save the frustration of trying to find it.

Managing the geodatabase schema

The structure of a geodatabase, including its feature datasets, feature classes, and tables, is known as the *schema*. Designing a geodatabase involves planning this structure and developing the datasets to match. Even in a well-designed database, however, the schema will sometimes need to be modified. For this reason, ArcGIS Pro designates these types of changes as *design changes*.

Most design or schema changes, such as creating a new feature dataset or feature class, importing or exporting feature classes, or adding and deleting fields in a table, operate in much the same way as ArcMap. However, ArcGIS Pro has some significant differences in how the structure of a table is modified.

Before launching into an explanation, however, it is important to clearly define three different types of changes that can be made to tables. *Properties changes* include operations such as sorting records, freezing fields, or changing the display format, and they affect only the properties of a specific table view in a specific map. *Editing changes* modify the content of tables and are stored in the source table, affecting every table view. *Schema changes* include operations that modify the underlying structure of the source table, including deleting fields, adding new fields, or specifying domains for a field.

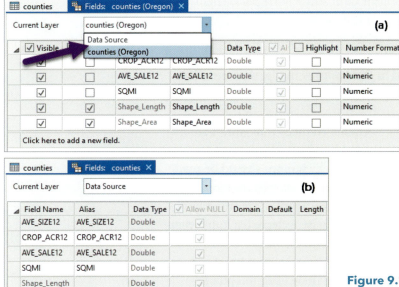

Figure 9.1. The Fields view is used to modify a table's (a) properties or (b) schema.

In ArcMap, it was usually clear which types of changes were being made. Properties changes were made by opening the table or layer properties. Editing changes had to be done within an edit session. Schema changes were mostly performed in ArcCatalog or the Catalog window. In ArcGIS Pro, the different types of changes are not so clearly delineated in the interface, so users must be extra careful to learn and remember when different types of changes are being performed.

When changes to a table are desired, the user will normally open the Fields view (figure 9.1), which may be accessed from the ribbon using the Table: View contextual tab or the Feature Layer: Data contextual tab. Additional buttons, such as the Add button in the table view, will also open the Fields view.

The Fields view has two modes controlled by the Current Layer drop-down button at the top of the view. When the mode is set to the layer/map name, counties (Oregon) in figure 9.1a, changes made to the fields are properties changes—i.e., they affect only the layer properties and not the source data. Hence, if an alias is changed in the counties (Oregon) mode, the new alias will appear in the counties layer in the Oregon map, but not in other layers, maps, or projects. However, adding a field is always a schema change, regardless of where it is done. If this counties (Oregon) mode is used to add a field, it will be added to the source data.

If the drop-down menu is changed to Data Source (figure 9.1b), edits will affect the stored source data and will be permanent. In the Data Source mode, a slightly different group of settings is presented: for example, the Visible check box and the Number Format settings are omitted because they do not apply to a source table. If the alias is changed in Data Source mode, the new alias will appear in all layers, views, and maps that reference the source data.

ArcGIS Pro offers additional schema changes that were not possible in ArcMap. It allows the field name itself, not just the field alias, to be renamed. However, such changes should be made sparingly. Field name changes, for example, may cause symbolized maps in this project or others to no longer work when the original field name has been changed. It is also wise to have backups whenever schema changes are being implemented.

When the Fields view is open, an additional contextual tab is available, the Fields tab (figure 9.2). It can be used to open the Domains or Subtypes editor, but the most important button is the Save button, in the Changes group. Changes in the Fields view will not take effect until the Save button is clicked. Both properties and schema changes must be saved. To cancel pending changes, close the Fields view and confirm the action to close the pane.

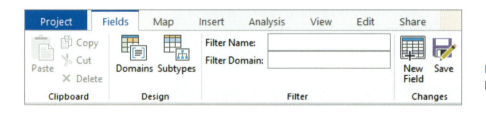

Figure 9.2. The Fields tab.

TIP *It is possible to open a different tab during changes in the Fields view, which can hide the Save button. Other actions may not be possible until the changes are saved, which is potentially confusing and frustrating. Make it a habit to save Fields changes immediately and close the Fields view before beginning another task.*

Creating domains

Domains, as part of the geodatabase data model, work the same way as domains in ArcMap, but the interface for manipulating them has changed. Because domains pertain to a geodatabase rather than an individual table or feature class, ArcCatalog was generally used to create or modify domains. In ArcGIS Pro, this function is accessible directly from several locations, including the Feature Layer: Data tab, the Fields tab, and the Fields view. Clicking any of these buttons opens the Domains view, and a Domains contextual tab is also opened on the ribbon (figure 9.3a). In figure 9.3b, the Domains view is open for a project named TablesPractice, and a coded value domain named PartyStatus is being created to characterize an object, such as a state, as Mostly Democratic, Mostly Republican, or Mixed Party.

The green cells, visible on the left of the domain name and the domain codes, are a signal that the edits being made are correct and can be saved. If one or more of the cells turns red, it signals that an error has been made, and the changes cannot be saved until the errors have been fixed. A similar strategy is also used when making changes in the Fields view.

Once created, domains created in ArcGIS Pro will work perfectly in ArcMap, and vice versa.

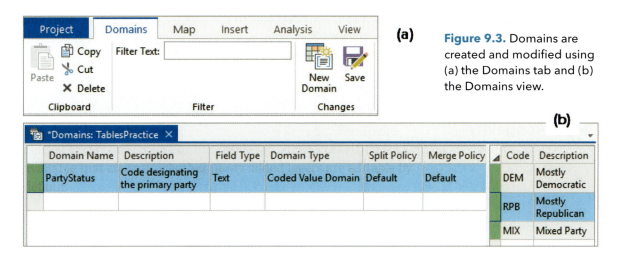

Figure 9.3. Domains are created and modified using (a) the Domains tab and (b) the Domains view.

TIP *It is possible to open a different tab during changes in the Domains view, which can hide the Save button. Other actions may not be possible until the changes are saved, which is potentially confusing and frustrating. Make it a habit to save domain changes immediately and close the Domains view before beginning another task.*

Managing data from diverse sources

The project data structure complicates general data management across an organization's or a user's broad data holdings. The project is designed to keep all the datasets related to a specific geographic area for an intended purpose, with folder links meant to access only data related to the project. Moreover, you start ArcGIS Pro by either opening or creating a project. A new project includes folders, a geodatabase, toolboxes, and so on, all created on the computer. This procedure becomes problematic when the aim is to manage information in a data repository that encompasses many geographic areas. Creating a whole new project each time does not make sense. ArcCatalog did not require you to open any data file, and in ArcMap you could open an untitled map document, do whatever data management was needed, and exit without saving a permanent file. Both programs also tracked all your data connections all the time, whereas in ArcGIS Pro they are independent for each project. This new way of managing data is an advantage most of the time, because it speeds data access and makes it easier to find the relevant folders and data while working on a project. However, it is not as convenient when managing large collections of data used by many projects.

One reasonable solution to this dilemma involves creating a specific project used whenever cross-project data management is needed. Give it a name, such as ManageData, and place in it all the folder connections that you commonly use. This approach allows quick access to multiple data collections and speeds data management tasks not associated with a specific project.

Project longevity

As the default output location for all geoprocessing tools and commands, the project geodatabase will, as time passes, acquire a vast collection of datasets associated with the project. Many of these datasets are likely to be temporary steps in a long analysis procedure, or even mistakes. Others will be important resources. Layers in multiple maps will refer to both types. Cleaning the geodatabase by deleting old and useless files will break layer links and require additional cleanup of maps.

Eventually, one must ask the question, "What is the useful life of a project?" Does it finally get so full that finding the data you want among the temporary and garbage files becomes a problem? Do you start over with a new project? If so, what about the datasets and maps and layouts that must be transferred for the new project to use? Although resources exist to copy datasets, layouts, models, and so on to a new project, the process could be tedious.

Projects are also a relatively new format for a data model, and it is not yet established whether they remain robust over the long term or whether, as occasionally happened with ArcMap map documents, they can become corrupted and no longer work properly. Again, manually transferring the critical elements to a new project is possible but tedious and without guarantee that all the important pieces can be salvaged. It seems foolhardy to assume that a project will work forever and summarily load it with five years' worth of data, maps, and layouts. However, continually creating new projects may reduce easy access to the prior work, or else make it more difficult to find a desired layout or dataset six months down the road because you can't remember in which version of the project it was saved. This question is one that all users will need to grapple with to find a balance that works for them.

Managing shared data for work groups

When a group of people are working on the same venture, the group members often need to share data, maps, layouts, and other items. Creating multiple copies of items, so that each person has his or her own copy, is the safest approach, because it reduces the possibility of corrupting files because of simultaneous access by different users. However, it complicates the difficulty of ensuring the consistency of the datasets. Many organizations do not have the capability or resources to run an enterprise geodatabase, which stores GIS data in a commercial database and supports multiuser editing. Unless one is available, the work group must develop a strategy to manage the shared data in a way that supports group members' work style and objectives. This issue is just as critical for ArcGIS Pro users as it has been for ArcMap users.

When developing sharing strategies for work groups, users should treat a project the same way as a map document or geodatabase for ArcMap. Either each user makes his or her own copy, or some procedure is used to guarantee that simultaneous use, especially editing, never occurs. Because projects accumulate items over time, it is best to assume that a project is intended to be used by an individual, not a group, unless special precautions are taken.

Because projects can access data through folder connections in addition to data stored within the project geodatabase, it makes sense to put shared datasets in external locations on file servers (appropriately protected by suitable read/write permissions) and use project geodatabases to store data used by individuals on their own computers. Users creating new datasets intended to be shared with others can keep the dataset within their own project until the dataset is complete, and then copy the dataset to the shared file server. This approach generally worked well for ArcMap and should continue to work for ArcGIS Pro.

ArcGIS Pro has the capability to package different elements of a project (layers, maps, layouts, even entire projects) into a single file that can be shared through a server or in ArcGIS Online. This method creates an easy way to share items between users (although they remain separate copies once they are copied or downloaded to a new individual).

TIP Any project can be saved as a project template, enabling the project, with all its folder connections, maps, and layouts, to be shared with others. Templates speed the setup of a new project while simultaneously standardizing the maps and layouts used.

Time to explore

Objective 9.1: Creating a project and exporting data to it

Creating datasets, importing and exporting feature classes, and other common data compilation tasks are easily learned in ArcGIS Pro. This time when you create a new project, you will use the Map system template so that the new project opens with a map already created.

1. **If ArcGIS Pro is already open, use the Quick Access Toolbar at the top to click the button 🖼 to create a new project. If not, start ArcGIS Pro and use the middle section to create a new project from the Map system template.**

2. **Name the project** EdwardsAquifer **and save it to the SwitchToProData folder (figure 9.4).**

Figure 9.4. Creating a new project.

3. **Rename the initial map view** Practice. Next, take a moment to set the map coordinate system to an appropriate projection.

4. **In the Contents pane, double-click the Practice map icon to open the map properties.**

5. **Click the Coordinate Systems settings.**

6. **Find the NAD 1983 StatePlane Texas Central FIPS 4203 (Meters) coordinate system and choose it. (Also click the star to add it to the Favorites tab.) Click OK.**

    Next, you'll find a dataset or two in ArcGIS Online and import them to the EdwardsAquifer home geodatabase. When looking for data, note that feature layers and layer packages will be the most likely to have exportable data.

7. **Open the Catalog view (not the Catalog pane).**

8. **In the Contents pane, showing the contents of the project, click the All Portal entry.**

9. **Use the Search box to search for datasets in** Austin, Texas. **Find a hosted feature layer or a layer package that looks interesting. Right-click and add it to the current map.**

10. **Switch back to the Practice map. If the added service is a group layer, expand it so that the individual layers are visible.** Only individual feature layers can be exported.

11. **In the map, right-click the feature layer and see if the menu includes the entry Data > Export features.**

12. **If the Export Features entry is there, the service is configured to allow a user to download and save a copy. If your service layer does not permit this, remove the service layer and try another.**

TIP Exporting a layer is not required to complete the exercises. If having trouble finding a service to export, just skip ahead to objective 9.2.

13. **Right-click the layer and click Data > Export Features. The Copy Features tool opens.**

14. **Edit the Output Feature Class name to make it clear and simple. Note that it will save the file to the project geodatabase by default. Don't run the tool yet.**

15. **Switch to the Environments tab on the tool.**

16. **Click the drop-down arrow for the Output Coordinate System and set it to the current map.**

17. **Run the tool.**

18. **Remove the hosted feature layer, leaving only the new feature class in the map.**

Objective 9.2: Creating feature classes

The SwitchToProData folder contains a map image that you'll be using, but you must add a folder connection to the folder to see it.

1. **On the ribbon, open the Insert tab and click the Add Folder button.**

2. **Navigate to and specify the SwitchToProData folder.**

3. **In the Catalog pane, expand the Folders entry and the SwitchToProData folder.**

4. **From the Catalog pane, drag the EdwardsCrop.png image to the map.**

5.   Right-click the EdwardsCrop layer and click Zoom To Layer.

Next, you will create three new feature classes to use when you learn to edit: a point feature class named karstfeatures, a line feature class named faults, and a polygon feature class named geology. You must make sure that each one is assigned the correct state plane coordinate system.

6.   In the Geoprocessing pane, search for the Create Feature Class tool.

7.   Enter the name karstfeatures **and set Geometry Type to Point (figure 9.5).**

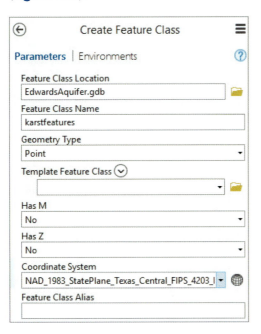

Figure 9.5. Creating a feature class for karst features.

8.   Click the Coordinate System drop-down arrow and set it to Current Map [Practice] or to the EdwardsCrop image.

9.   Run the tool and leave it open.

TIP *Leaving the tool open lets you change a couple of settings and run it again quickly.*

10. Change the feature class name to faults **and the geometry type to** Polyline **and run the tool again.**

11. **Change the feature class name to** geology **and the geometry type to** Polygon **and run the tool a third time.**

12. **Save the EdwardsAquifer project.**

Objective 9.3: Creating and managing metadata

Editing metadata in ArcGIS Pro is not that different from ArcMap. Different metadata styles are available, with the streamlined Item Description as the default.

1. **On the ribbon, open the Project tab and click Options. Open the Metadata section and view the settings.**

2. **Select your customary metadata style or leave it set to the Item Description style.**

3. **Click OK to close the options and use the back arrow to return to the project.**

4. **Switch to the Catalog view and navigate inside the Folders heading to the project folder and inside the project geodatabase, EdwardsAquifer.gdb.** Four feature classes should appear in the Catalog view: faults, geology, karstfeatures, and the feature class you saved from ArcGIS Online (figure 9.6a).

5. **Make sure that the faults feature class is selected in the Catalog view.**

6. **Open the Home tab and, in the Metadata group, click Edit.**

7. **Fill out the information shown in figure 9.6b, and then click Save in the Manage Metadata group on the Metadata contextual tab.**

8. **Close the faults metadata window.** In ArcGIS Pro, the metadata stored with the source feature class is automatically shown in a map layer. If you want to have different metadata for the layer, you can change the setting to create other metadata for just the layer, either copying the source metadata for editing or creating it from scratch.

9. **Open the Practice map.**

10. **Open the faults layer properties and view the Metadata section.**

11. **To create separate metadata for the layer, change the top drop-down arrow to Layer has its own metadata, and if you want, click the button to copy the source metadata to the layer.**

12. **Experiment with this window for a moment. It doesn't matter which type of metadata you choose for the layer right now.**

TIP *Remember that the layer metadata is separate and distinct from the source metadata stored with the dataset (just as the layer title may be different from the dataset name). If you want to edit the source metadata that is a permanent part of the dataset, use the Catalog view.*

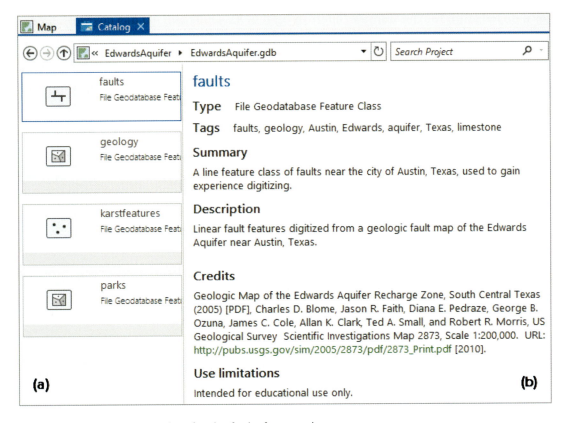

Figure 9.6. Creating metadata for the faults feature class.

Objective 9.4: Creating fields and domains

Now you will set up the tables for the karst features, faults, and geology feature classes. These feature classes need fields to store the attributes, including coded domains for the fault type and the karst features. Domains are created for the geodatabase, so the field does not need to exist for a domain to be added. You will create the domains first, and then associate them with fields.

1. **Switch to the Catalog view. Navigate to (not inside) the EdwardsAquifer.gdb geodatabase.**

2. **Right-click EdwardsAquifer.gdb and click Domains.**

TIP You may see some domains already if the dataset you imported from ArcGIS Online had them. Just ignore them.

3. **Click in the first empty Domain Name box and enter** KarstType.

4. **Enter a description for the domain, such as** Type of karst feature.

5. **Let Field Type default to Text and Domain Type default to Coded Value Domain.**

6. **Click in the first Code box and, for both Code and Description, enter** spring.

7. **In the next row, enter** cave/sink **for both. Make sure not to press the Tab key! Click on the first row to establish the last entry and signal that you are finished entering codes.**

TIP Pressing the Tab key starts a new code row, and the domain cannot be saved until values are entered. If you press Tab by mistake, select the blank row and delete it. Red cells next to the domain rows indicate that an error must be fixed before the domains can be saved.

8. **For Domain Name, click in the next empty box and enter** SymbolType. **Also add a description such as** Type of line symbols.

9. **Set the Field Type box to Text. Domain Type will default to Coded Value Domain.**

10. **Click in the first Code box and, for both Code and Description, enter** solid**. In the second row, type** dashed **for both, and do** *not* **press the Tab key.** The domains should appear as in figure 9.7. The Code/Description section will show the codes for whichever domain is highlighted.

11. **Click Save and close the Domains view.**

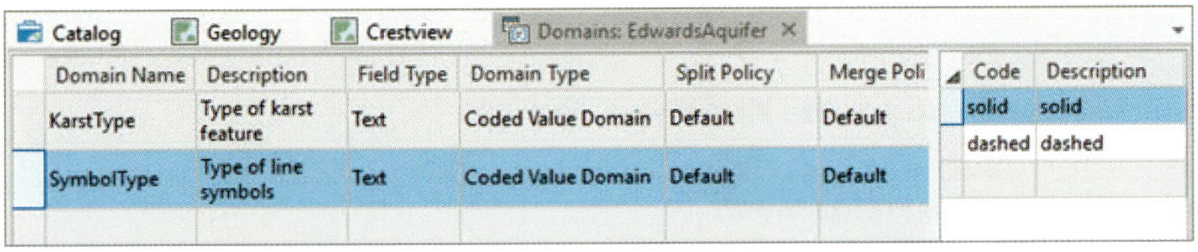

Figure 9.7.
Domains for the EdwardsAquifer geodatabase.

Next, you'll add the fields to the fault and geology feature classes, taking care to assign the domain.

12. **Open the Practice map.**

13. **In Contents, click faults to select it, and on the Feature Layer: Data contextual tab, in the Design group, click Fields to open the Fields view.**

14. **Make sure that the Fields view is wide enough that the Length column is visible, or scroll to it as needed.**

15. **Click to add a new field and, for Field Name and Alias, type** Linetype**.**

16. **Click inside the Data Type box and choose Text. Set Length to** 10**.**

17. **Click in the Domain box and choose the SymbolType domain.**

18. **On the Fields tab, in the Changes group, click Save and close the Fields view.**

TIP *Remember that it is possible to open a different tab during changes in the Fields view, which can hide the Save button. Other actions may not be possible until the changes are saved, which is potentially confusing and frustrating. Make it a habit to save the field changes immediately and close the Fields view before beginning another task.*

19. **In Contents, click on the karstfeatures layer and open its Fields view.**

20. **Add a new field named** Type, **making it a Text field with a length of** 15**.**

21. **Click in the Domain box and choose the KarstType domain.**

22. **Save the changes and close the Fields view.**

23. **Open the Fields view for the geology layer and add a text field named** Unit **with a length of** 6**. No domain is needed for this field.**

24. **Save the changes and close the Fields view.**

25. **Save the EdwardsAquifer project.**

Objective 9.5: Modifying the table schema

One of the perks of ArcGIS Pro is that it allows table schema modifications that were not possible in ArcMap. Next, you'll rename a field and reorder the fields in the faults table.

1. **Open the Fields view for faults (figure 9.8).**

2. **Change both Field Name and Alias for the Linetype field to** FaultSymbol**.**

3. **Click the green cell on the left of the FaultSymbol field name and drag the row above the Shape field.**

4. **Save the changes and close the Fields view.**

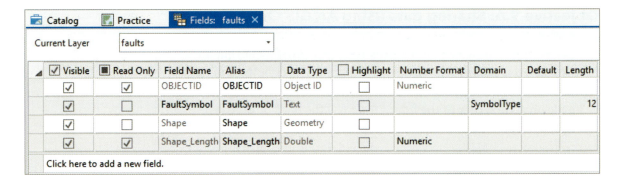

	▣ Visible	▣ Read Only	Field Name	Alias	Data Type	☐ Highlight	Number Format	Domain	Default	Length
	☑	☑	OBJECTID	OBJECTID	Object ID	☐	Numeric			
	☑	☐	FaultSymbol	FaultSymbol	Text	☐		SymbolType		12
	☑	☐	Shape	Shape	Geometry	☐				
	☑	☑	Shape_Length	Shape_Length	Double	☐	Numeric			

Click here to add a new field.

Catalog   Practice   Fields: faults ✕

Current Layer   faults ▾

Figure 9.8. Modifying the table view properties.

5. **Open the faults attribute table and examine the fields.**

6. **Add another copy of faults from the EdwardsAquifer geodatabase and rename it** faults2.

7. **Open the faults2 table and compare it with the faults table.** When you made the change, the Fields view was set to the layer properties mode. The alias change affected only the layer, not the data source. To change the alias permanently, the Data Source mode is used.

8. **Remove the faults2 layer.**

9. **Reopen the Fields view for faults.**

10. **In the Fields view, change the Current Layer drop-down box to Data Source. Notice that the field name was changed to FaultSymbol, but the alias is still Linetype.**

11. **Change the alias to** FaultSymbol **so it will appear in all maps that way.**

12. **Change the length of the FaultSymbol field to** 12.

13. **Try to change the order of the fields as you did before.** It is not possible. Reordering fields is a cosmetic property change, not a schema change.

14. **Save the changes and close the Fields view.**

15. **Save the EdwardsAquifer project.**

Objective 9.6: Sharing data using ArcGIS Online

ArcGIS Pro makes it easy to share information with others. Imagine that you have set up this EdwardsAquifer project but that a GIS intern is going to do the actual digitizing. By creating a project package, you can transfer the entire setup, either by saving it as a file on a file server (if the intern is across the hall) or in ArcGIS Online (if the intern works in another office location).

1.   **On the ribbon, open the Share tab and, in the Package group, choose Project.**

2.   **In the Package Project pane, under Start Packaging, click Upload package to Online account (figure 9.9).**

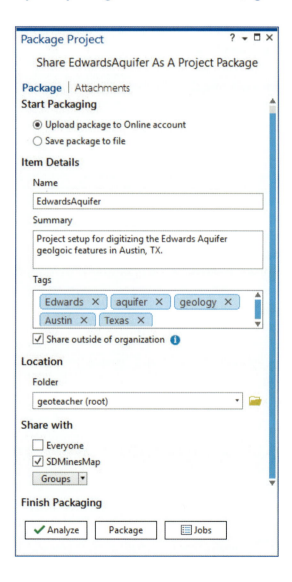

Figure 9.9. Packaging a project.

TIP *If you don't have an ArcGIS Online account, choose Save package to file instead and save it to a location on your hard drive.*

3. **Fill in the Summary and Tags boxes for the package contents.**

4. **Select a folder in your ArcGIS Online My Content holdings (or on your hard drive) to store the package in.**

5. **Click the blue circled *i* next to Share outside of organization and read the information. Check the box (we'll assume that the intern is in another office).**

6. **You want the intern to be able to see the package, so under Share with, check the box for the name of your organization (SDMinesMap in figure 9.9, but yours will be different).**

7. **Click Analyze to check the package for errors.**

8. **If no errors are found, click Package to finish packing the project.**

9. **Close the Package Project pane.**

10. **Save the EdwardsAquifer project.**

TIP *You can also package individual layers or maps if you want to share just portions of a project with others. This method is similar to packaging projects.*

Chapter 10

Editing

Background

Editing has changed significantly in ArcGIS Pro, although some
of the methods, such as using feature templates, remain similar.
For the most part, users will find the new interface more intuitive
and streamlined than ArcMap. We will begin with a review of the
editing tools and panes, and then we will examine how to create
features, modify them, and edit annotation.

Basic editing functions

One of the largest changes in ArcGIS Pro is that edit sessions no
longer exist. All layers are editable by default and can be edited at
any time. If this news gives you the willies and visions of accidentally damaged datasets, don't panic. You can turn off editing
for some or all layers as part of the project, to protect them from
accidental changes.

The Contents pane contains two graphical tabs that are especially useful when editing. The List By Editing tab (figure 10.1a)
has check boxes that control whether layers are editable. To dis-
able editing for a layer, clear the check box. A project setting is also
available that controls whether newly added layers are editable by
default.

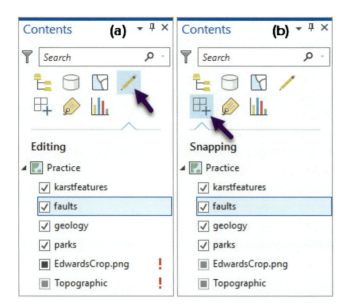

Figure 10.1. Tabs in the Contents pane that can be used for editing include (a) the List By Editing tab and (b) the List By Snapping tab.

The List By Snapping tab of the Contents pane (figure 10.1b) controls which layers have snapping enabled. The order in the list also determines the snapping precedence: if multiple layers have features close together, the highest one in the list will be preferentially snapped. Snapping is similar in ArcGIS Pro in other ways as well, with the addition of a convenient pop-up Snapping menu underneath the display window.

The Edit tab on the ribbon accesses most of the editing tools and functions (figure 10.2). The Features group opens the editing tools for creating or modifying features. The Selection group matches the group by the same name on the Map tab; ArcGIS Pro does not have a special selection tool for editing as in ArcMap. ArcGIS Pro supports editing with no topology, the simple map topology, or an established planar topology.

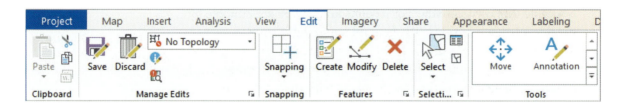

Figure 10.2. The Edit tab.

Creating features

ArcGIS Pro uses feature templates to create new features, although these templates have some new capabilities and are quicker and easier to use than those in ArcMap. You can open the Create Features pane using the Create button on the Edit tab, in the Features group. The main mode for creating features, the Create Features pane (figure 10.3a), lists all available templates in the map. When a template is clicked, such as geology, a blue arrow appears that will open the Active Template mode (figure 10.3b). This mode allows attribute features to be specified before the feature is added.

As in ArcMap, a variety of different construction tools are available for each geometry type, including new ones not available in ArcMap. Stream digitizing, however, has not been carried into ArcGIS Pro; users must content themselves with the somewhat similar Freehand construction tool.

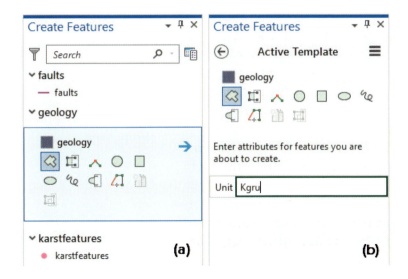

Figure 10.3. A user creates features with: (a) the Create Features pane with its feature templates and (b) the Active Template mode.

As with ArcMap, construction menus are available to access specific editing functions. The general menu (figure 10.4a) appears when the map is right-clicked and no construction tool is active. When sketching, right-clicking in an empty area off the sketch opens the Sketch menu (figure 10.4b). As with the ArcMap menu, it provides segment constraints used to specify the next vertex, such as making it appear at a specific distance or follow a given direction. However, additional options are now available to specify directions. In ArcMap, east was given the direction of 0, north was 90, and so on. In ArcGIS Pro, several options for specifying directions are available, with the default method using bearings—e.g., N20E or S15W.

ArcGIS Pro also has a new feature, called *dynamic constraints*, which facilitates editing when segment constraints are used frequently. Dynamic constraints continuously show distances and bearings as features are added (figure 10.4c).

The familiar Vertex menu (figure 10.4d), which appears when the user right-clicks on a sketch, is also similar to the ArcMap version.

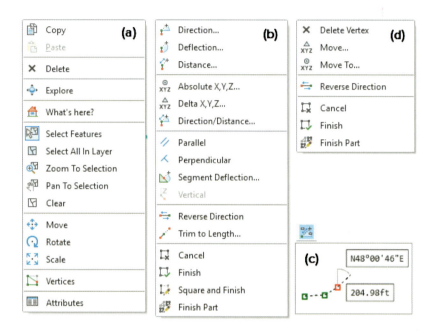

Figure 10.4. Editing context menus: (a) the general menu, (b) the Sketch menu, (c) dynamic constraints, and (d) the Vertex menu.

Modifying existing features

The tools for modifying existing features have been collected into a single editing pane, the Modify Features pane (figure 10.5a), opened by clicking Modify on the Edit tab, in the Features group. The pane includes multiple headings with tools for different tasks: aligning, reshaping, dividing, and so on.

When a tool, such as the Reshape tool, is clicked, the tool opens in the pane (figure 10.5b). If no feature was selected before opening the tool, the user will be prompted to make a selection; you can also change the selection from within the tool (ArcMap editors will appreciate this capability).

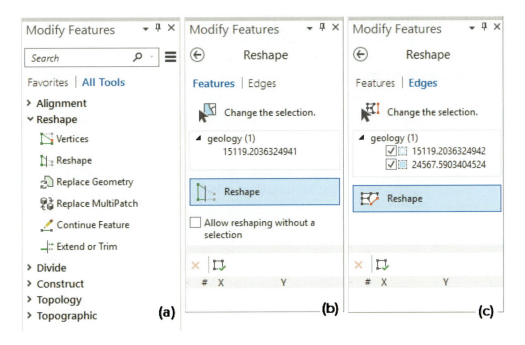

Figure 10.5. The Modify Features pane contains tools for (a) modifying features, (b) reshaping features without using topology, and (c) reshaping edges with map topology.

Topological editing is also supported in ArcGIS Pro. By default, no topology is available during editing, but a simple map topology can be built and used instantly by changing the Topology drop-down arrow on the Edit tab, in the Manage Edits group (see figure 10.2).

When a topology is present, either a temporary map topology or a permanent planar topology, some Modify Features tools possess an Edges tab (figure 10.5c) that can be used to edit edges or nodes of a topology, such as when modifying the boundary between two adjacent polygons. This function replaces the Topology Edit tool and the Topology Toolbar functions that implemented this capability in ArcMap.

TIP *You may find it convenient to dock the Create Features and Modify Features panes on top of each other in the same window and switch back and forth while editing.*

Creating and editing annotation

ArcGIS Pro supports only the type of annotation that is stored in a geodatabase, either as standard or feature-linked annotation. Creating annotation that can be stored in a map (data frame) and edited with the Draw toolbar is no longer possible. As such, annotation can only be edited using the regular editing functions of ArcGIS Pro.

Annotation can be created one item at a time using the Create Features pane, or it can be created from dynamic labels that have been set up for a layer. This approach is the same as the ArcMap strategy, except that the menu window for creating annotation has been replaced by the Convert Labels to Annotation geoprocessing tool.

Once created, geodatabase annotation can be edited the same as any other features. Various construction tools are used to create new annotation or modify existing annotation, and the familiar Attributes pane is also available. The earlier versions of ArcGIS Pro had only one basic construction tool for creating Straight annotation, but additional tools are being added as new versions debut. Editing the path of curved annotation, placing annotation that automatically follows a feature, and the follow feature construction tool are not supported in version 2.3. Leader line annotation can be edited but the construction tool is not yet available. The placement of unplaced annotation is handled manually now without an overflow window, requiring editing of the attributes to convert it to placed annotation. Users who rely on certain advanced strategies might need to stick with ArcMap for editing annotation, at least for a while.

Time to explore

Objective 10.1: Understanding the editing tools in ArcGIS Pro

We will begin with a tour of the various windows and settings for editing to get familiar with the interface.

1. **Open the EdwardsAquifer project and the Practice map, showing the karstfeatures, faults, and geology feature classes created in chapter 9.**

2. **In the Contents pane, click the List By Editing tab** ✏. This tab is used to control which layers may be edited. To protect a layer from accidental changes, clear the check box next to its name. The Topographic basemap and EdwardsCrop image have red exclamation marks (!) because they are rasters and cannot be edited.

3. **Clear the check box for your downloaded layer, if present, but leave the others checked.**

4. **In the Contents pane, click the List By Snapping tab** ⊞₊**.** The check boxes on this tab turn snapping on and off for each layer. The order of layers, from top to bottom, also specifies the snapping precedence.

5. **Clear the check box for your downloaded layer.**

6. **Drag the geology layer to the top of the list to give it precedence.** Take a moment to look at how the snapping compares with ArcMap. Snapping options are accessible on the Edit tab, in the Snapping group, and from a pop-up Snapping menu underneath the map.

7. **On the ribbon, open the Edit tab, and in the Snapping group, click the Snapping drop-down arrow.**

8. **Click the icon that states "Snapping is On" (or Off) to toggle snapping availability.**

9. **Examine the snapping types in the lower row of the Snapping drop-down menu, which are the same as for ArcMap.**

10. **Open Snapping Settings and examine the settings. Close the dialog box when done.**

11. **Look for the small snapping icon** ⊞₊ **on the right of the scale readout on the bottom edge of the map window.** It will be blue if snapping is on and white if snapping is off.

12. **Point to it to open the same snapping options.**

13. **Examine the Edit tab groups and buttons.** Most of the editing functions are accessed from the Edit tab on the ribbon.

14. **In the Selection group, click the Attributes button ▦ to open the Attributes pane.**

15. **Dock the Attributes pane on top of the window containing the Catalog pane for easy access.**

    Finally, let's take a look at the project editing options, which contain many useful customizations.

16. **Click Project > Options and choose the Editing settings. Examine them in detail.** Note the option under Settings to automatically save edits after a specified period of time or number of edits.

17. **Click the back arrow to return to the map.**

Objective 10.2: Creating points

To fully explore editing, you'll need to create some features to work with, using your three new feature classes.

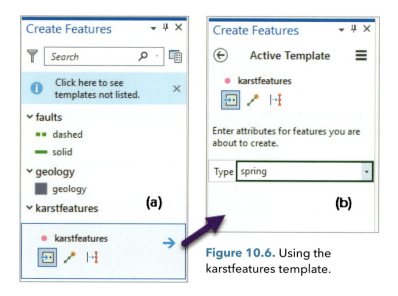

Figure 10.6. Using the karstfeatures template.

1. **On the Edit tab on the ribbon, in the Features group, click the Create button to open the Create Features pane (figure 10.6a). Dock it in the same window as the Catalog and Attributes panes.**

2. **Click the karstfeatures template and notice the three construction tools available. You will use the default Point construction tool on the left.**

3. **Click the blue arrow on the karstfeatures template to open the Active Template mode (figure 10.6b).**

4. **Click in the Type field box and choose the spring entry (available from the domain created in objective 9.4 of chapter 9).**

Several shortcuts assist in zooming/panning while using a template, rather than returning to the Explore tool on the Map tab. These shortcuts allow you to zoom or pan without interrupting a sketch. Now you can try them out.

5. **Rotate the mouse wheel to zoom in or out using the preset scales.**
   a. **For a continuous zoom, right-click on the map and move the pointer up or down.**
   b. **To pan, click the mouse wheel and drag.**

6. **Experiment with these techniques until they are comfortable.**

TIP If you like using editing shortcuts, search ArcGIS Pro Help for "editing shortcuts" to learn more about how they work in ArcGIS Pro.

Okay, now you can try entering some karst points.

7. **Find and zoom to some blue dots on the map. Click in their centers to enter points.**

8. **In the Active Template mode, change the Type field box to cave/sink.**

9. **Find and zoom to some black dots on the map. Click in their centers to enter points.**

Objective 10.3: Creating lines

Now you'll enter some faults.

1. **In the Active Template mode, click the back arrow to return to the Create Features pane.**

2. **Click the faults template. Examine the construction tools but use the default line tool on the left.**

3. **Click the blue arrow to open the Active Template mode.**

4. **Set the FaultSymbol attribute field to dashed.**

5. **Click the map to enter several dashed faults, double-clicking to end each one.**

6. **Change the FaultSymbol attribute field to solid.**

7. **Use the Snapping drop-down button on the Edit tab or the pop-up Snapping menu  on the status bar below the map to make sure that snapping is on.**

8. **Enter several solid faults near each other. Pay attention to the snap tips as they appear.**

9. **Use the pop-up Snapping menu at the bottom edge of the map to turn different types of snapping on/off as needed.**

10. **On the Edit tab, in the Manage Edits group, click the Save button to save the edits.**

Note the Construction toolbar that appears near the bottom of the map when entering lines or polygons (figure 10.7). It replaces some of the tool buttons on the ArcMap Edit toolbar, which provided access to different sketching tools, including, from left, the straight-line segment, the right-angle corner tool, curves, and tracing. The check box and X box are used to finish and cancel a sketch, respectively.

Figure 10.7. The Construction toolbar.

11. **Zoom to a clear area of the map.**

12. **Add some practice lines using the different tools, to get accustomed to them.**

13. **On the Edit tab, in the Selection group, use the Select button to select the practice lines and, in the Features group, use the Delete button to delete them. The keyboard Delete key may also be used to delete features.**

14. **Save the edits.**

    The fault symbols are hard to see against the black of the image. You can symbolize them by fault type with thicker, contrasting lines, which has the added benefit of creating templates for each fault type.

15. **Open the Symbology pane using the Feature Layer: Appearance tab or by right-clicking the faults layer in the Contents pane or by clicking the Symbology tab if the pane is still docked in a window of the GUI.**

16. **Make sure that the faults layer is selected in the Contents pane.**

17. **Change the Symbology type to Unique Values and use the FaultSymbol field.**

18. **Click the Add All Values button to populate the Unique Values list.**

19. **Click the symbol for the dashed class to modify it. Find a dashed symbol on the Gallery tab, and then switch to Properties to make the symbol thick and a distinctive color.**

20. **Repeat for the solid line symbol.**

21. **Close the Symbology pane when finished.**

22. **Return to the Create Features pane.** You should now see two templates under faults, one for solid and one for dashed, matching the new symbols. The attributes in the FaultSymbol field will continue to be entered in new features as you use these templates.

23. **Add another dashed line using the template. While it is still selected, switch to the Attributes pane and confirm that the FaultSymbol attribute has been entered.** The Attributes pane is like the Attributes window in ArcMap. However, it has two tabs, Attributes and Geometry, in the lower half of the pane. The Geometry tab replaces the ArcMap Show Geometry tool.

TIP If you dock the editing panes in a single window, it makes it easy to switch back and forth between them as needed, while conserving screen real estate for editing (figure 10.8).

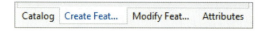

Figure 10.8. Docking the editing panes in one window for easy access. The tabs for each pane show up at the bottom of the open pane.

Objective 10.4: Creating polygons

Finally, try creating some polygons.

1. **Zoom to the southwest corner of the map and examine the geology unit polygons.**

2. **In the Create Features pane, open the Active Template mode for the geology template.**

3. **Type Kkd in the Unit attribute box and digitize one of the hot-pink polygons. It's okay to be sloppy; this is just practice.**

FYI: Geologists use short codes to label rock units. The first uppercase letter indicates the geologic period, or age, of the rock, and the lowercase letters represent the formation name. Kkd stands for the Cretaceous-age dolomitic member of the Kalner Formation. The letter K is used for Cretaceous because C is used for Cambrian-age rocks.

Next, label the polygons with the geologic units.

4. **In Contents, click the geology layer, open the Feature Layer: Labeling tab, and set up labels for the geology polygons using the Unit field.**

5. **Return to the Edit tab.** Adjacent polygons should be digitized using the Autocomplete Polygon tool.

6. **In the Active Template mode, change the Unit attribute to Kkbn (the basal nodular member of the Kalner Formation).**

7. **Switch to the Autocomplete Polygon construction tool** ⬚.

8. **Digitize the purple polygon adjacent to the first one, making sure to snap to it and entering only the new boundary.**

9. **Change the attribute to Kgru (upper member of the Glen Rose Limestone) and digitize one of the adjacent brown polygons.**

   Finally, take some time to get acquainted with the editing context menus in ArcGIS Pro. They are similar to the same versions of these menus in ArcMap.

10. **Zoom to a clear area of the map.**

11. **Switch back to the Polygon construction tool in the geology template.**

12. **Enter a few vertices of a new polygon and right-click on the sketch boundary to open the Vertex menu (see figure 10.4d).**

13. **Add or delete a vertex.**

14. **Enter a few more vertices and right-click in an empty area off the sketch to open the Sketch menu (see figure 10.4b).**

15. **Use one of the sketching tools.**

16. **Spend a few minutes experimenting with these two menus to figure out the editing tools that you use most.**

17. **Use the Save button on the Edit tab, in the Manage Edits group, to save the edits.**

18. **Save the EdwardsAquifer project.**

TIP As in ArcMap, saving the project does not save the edits, and vice versa.

Objective 10.5: Modifying existing features

Creating features is only one part of editing. Modifying existing ones is also often needed. In ArcGIS Pro, all the modification tools are packed into one pane.

1. **On the Edit tab on the ribbon, in the Features group, click Modify to open the Modify Features pane. Dock it in the same window as the Create Features and Attribute panes.**

2. **Expand the headings and examine the tools. Many of them should appear familiar to you, at least the ones you used often in ArcMap.**

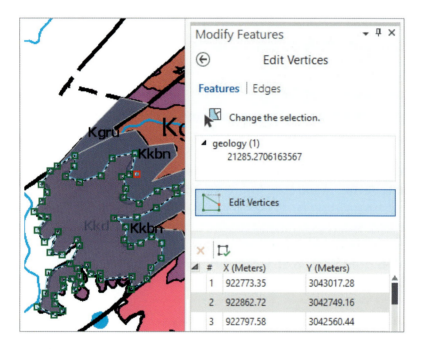

Figure 10.9. Using the Vertices tool to change the shape of a polygon.

You'll try using the Vertices tool to get familiar with how this pane works.

3. **Expand the Reshape heading and select the Vertices tool, which opens the Edit Vertices mode of the Modify Features pane.**

4. **Use the Select tool on the Edit tab, in the Selection group, or the Change the selection button in the Edit Vertices mode, to select one of the practice polygons made earlier in this chapter (figure 10.9).**

5. **Drag vertices to change the polygon shape. Right-click between vertices to open the Vertex menu and add vertices.** Notice the Construction toolbar near the bottom of the map. It contains slightly different tools from those in figure 10.7. These toolbars are sensitive to the operation being performed.

6. **Click the Finish tool ☑ on the Construction toolbar to finish editing vertices (or double-click the map).**

7. **Zoom to the area where you entered the three adjacent geology polygons.** To explore how topologic editing works, you'll need to turn on the map topology and work with some adjacent polygons.

8. **On the Edit tab on the ribbon, in the Manage Edits group, change the Topology drop-down box to Map Topology.** Notice that a second tab appeared in the Edit Vertices mode, named Edges.

9. **Switch to the Edges tab.**

10. **Use the selection tool in the Edit Vertices mode to select an edge between two of the polygons.**

11. **Modify the vertices of the edge and finish the sketch.** We can only touch the basics of editing in this lesson, but with this introduction, you should be able to work out most of your common editing tasks.

12. **Use the existing polygons and lines, or create new ones, and spend 10 or 15 minutes experimenting with the editing functions that you use the most.**

13. **Save the edits.**

14. **Save the EdwardsAquifer project.**

TIP It is possible to move to a different ribbon during editing, but certain functions may not be available if you have unsaved edits, particularly those that affect a feature class or table schema. Make it a habit to complete and save edits before beginning a new task.

Objective 10.6: Creating an annotation feature class

Not everyone uses annotation regularly, but those who do can get started using this exercise. The map-based annotation (which was the same as graphic text) is not available in ArcGIS Pro. All annotation must be created and stored as a geodatabase feature class. For this lesson, we return to the CraterLake project.

1. **Open the CraterLake project.**

2. **Close any open views. Use the Insert tab, and the Project Group, to create a new map. Name it** GeologyAnno.

3. **Add the geologyunits feature class from the CraterLake project geodatabase.**

   As in ArcMap, the first step in creating annotation from labels is to get the labels as close to the final result as possible. This step will be a good review of labeling from chapter 5, too.

4. **Zoom to** 1:10,000**, which will be a reasonable reference scale.**

5. **Select the geologyunits layer in the Contents pane and open the Feature Layer: Labeling tab on the ribbon.**

6. **Set the Field box to PTYPE.**

7. **Set the text symbol to Arial** 8-**pt regular black font.**

8. **Use the Label Placement Style drop-down arrow to select the Basic Polygon style.**

9. **Turn on the labels and examine them.** You'll need a few more tweaks. First, geology unit labels should always be horizontal on the page, and every label should be kept, even if the label extends outside the polygon.

10. **On the Labeling tab, click the Options dialog box launcher** ⌐ **in the lower-right corner of the Label Placement group. The Label Class pane will open.**

11. **Switch to the Position tab and expand the Placement heading.**

12. Change the placement to Regular placement and Horizontal in polygon (figure 10.10). Check the box May place label outside polygon boundary.

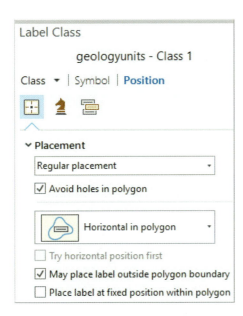

Figure 10.10. Setting the placement properties for the geology labels.

The geology labels should appear similar to those in figure 10.11a. Once the labels are the way you want them, it is time to create the annotation.

13. In the Geoprocessing pane, search for and open the Convert Labels To Annotation tool.

14. Fill out the tool as shown in figure 10.11b, making sure to save the annotation in the CraterLake geodatabase. Run the tool.

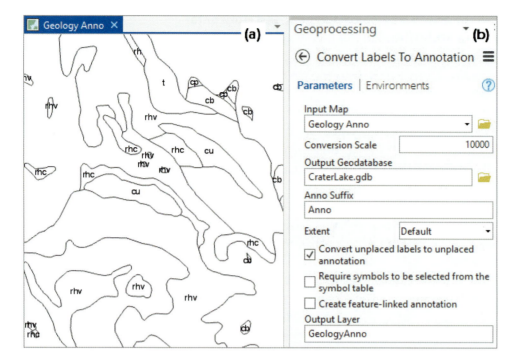

Figure 10.11. Labels are shown for (a) the geologyunits layer and (b) converting the geology labels to annotation.

Objective 10.7: Editing annotation

Editing annotation uses the same tools and panes as editing other features.

1. **Zoom to a block of annotation features that look as if they need editing.**

2. **Open the Edit tab on the ribbon and use it to open the Modify Features pane.**

3. **In the Modify Features pane, expand the Alignment heading and choose the Annotation tool.**

4. **Select a piece of annotation.** As in ArcMap, a selected annotation feature can be modified with the mouse in several ways. The pointer icon changes shape as the mouse is moved around the feature selection, indicating what will happen when dragging.

5. **Find the curved arrow icon and drag the mouse pointer to rotate the annotation.**

6. **Find the double-headed arrow and drag the pointer to increase or decrease the size.**

7. **Find the four-headed arrow and drag the pointer to move the annotation.**

8. **With the annotation still selected, open the Attributes pane by clicking the Attributes button ▦ on the Edit tab, in the Selection group.** As in ArcMap, the annotation properties can be edited in the Attributes pane.

9. **Examine the Annotation tab in the lower half of the pane, used to modify the symbol and other properties.**

10. **Switch to the Attributes tab and examine the fields.**

11. **Change the Status to Unplaced.** The annotation will disappear once it is no longer selected.

12. **Try selecting another annotation feature and move it to a better position.**

Objective 10.8: Creating annotation features

New annotation features can be created individually, but ArcGIS Pro does not yet have as many construction tools as ArcMap.

1. **Open the Create Features pane.**

2. **Choose the GeologyAnno template.**

3. **Use the default Horizontal construction tool, first on the left.**

4. **Type a word in the text box (Vent in figure 10.12).**

5. **Click on the map to place the word and create the annotation feature.**

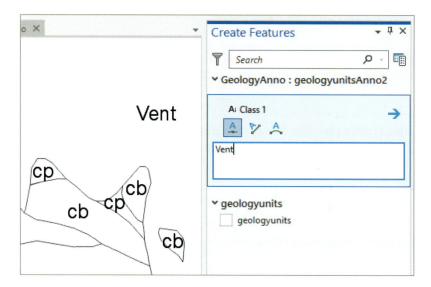

Figure 10.12. Creating a new annotation feature.

To create more varied annotation with different symbols, the Active Template mode is helpful.

6. **Click the blue arrow on the template to open the Active Template mode.**

7. **Change the font to a different size, color, or style.**

TIP The Active Template mode is used only to create new annotation features. To edit existing annotation, you must switch back to the Modify Features pane.

8. **Type a different word in the text box and create the annotation on the map.**

9. **Take a moment to explore the other construction tools for annotation.**

10. **Choose the Straight Annotation construction tool. Click twice to define a line along which the annotation will be placed.**

11. **Choose the Curved Annotation tool. Click to define a curve for the annotation. Double-click to finish, or right-click on the map and click Finish.**

You can return to the Modify Features pane to make changes to your new annotation features.

12. **Open the Modify Features pane and click the Annotation tool.**

13. **Click the button for Change the selection and select one of the annotation features.**

14. **Right-click the selected feature to open the annotation editing context menu. Examine the choices, which are like those in ArcMap.**

15. **With a piece of annotation selected, right-click a polygon feature edge and click Follow This Feature.**

16. **Play with the annotation construction and modification tools for a few more minutes.**

17. **Save the edits and save the CraterLake project.**

Chapter 11

Moving forward

This book covers the most basic features of the new ArcGIS Pro software. After getting familiar with the basics, or driven by a specific need, you will want to explore the additional capabilities that ArcGIS Pro offers, both those carried over from ArcMap and those that are new. The help is often the first place to look. The GeoNet website sponsored by Esri also helps you follow trends and learn new information, and it can be a great place to get your questions answered. Some advanced capabilities include the following:

- **Leveraging ArcGIS Online.** The tight integration of ArcGIS Pro with ArcGIS Online opens many opportunities for sharing data and workflows with coworkers or the public. Explore the types of services available for use.

- **Sharing feature layer, tile layer, and map services in ArcGIS Online.** Unlike creating packages, this type of sharing requires an ArcGIS Online organizational account configured with publisher permissions, so talk to your ArcGIS Online administrator first.

- **Learning to collect data in the field on mobile devices.** Several mobile apps support this activity, including ArcGIS Explorer, Collector for ArcGIS, and Survey123 for ArcGIS. This technique requires the ability to publish feature layer services, configure web maps, and design geodatabases with suitable fields and domains to foster efficient data gathering.

- **Styles.** ArcGIS Pro stores and manages symbol styles differently from ArcMap, and the old styles must be imported to be used. Learn to create and manage the new styles and symbols.

- **Map books and map series.** Building map series based on tiles and dynamic text continues to be a functionality available in ArcMap and supported by ArcGIS Pro.

- **Building planar (geodatabase) topology for data layers.** The map topology discussed in chapter 10 tracks only features that are adjacent or connected. A planar topology permits the user to specify rules about the spatial

relationships between features, test when the rules are vio-
lated, and fix these errors.

- **Geostatistical analysis.** The geostatistical analysis tools
available in ArcMap continue to be supported and expanded
in ArcGIS Pro, including Moran's I, hot spot analysis, krig-
ing, interpolation, regression, geographically weighted
regression, and other vital tools for exploring data and infer-
ring relationships.

- **Analyzing and editing 3D data.** Both geoprocess-
ing-based and interactive methods for 3D analysis are
supported, such as line of sight, view dome, and viewshed.
Objects such as multipatch features and x,y,z data can be
created, imported, or edited.

- **Time analysis and animation.** ArcGIS Pro includes and
expands the ArcMap capabilities to map and analyze information associated with a time field, including fly-throughs,
animation, and space-time data exploration using the
NetCDF data format.

- **Extensions.** ArcGIS Pro supports ArcMap extensions such
as ArcGIS Spatial Analyst™, ArcGIS Geostatistical Analyst, ArcGIS Network Analyst, ArcGIS Business Analyst™,
ArcGIS 3D Analyst™, and so on, extending the tools and
functions available to users.

- **Custom workflows.** The ability to create models in
ModelBuilder or write scripts in Python is still supported
in ArcGIS Pro. In addition, a new workflow sharing method,
the *task*, permits a workflow to lead a user through an interactive task with decisions and actions being taken by the
user along the way.

Data sources

Data was compiled from public sources, including the US Geological Survey (USGS) and Esri, by Maribeth H. Price. All datasets used from Data and Maps for ArcGIS were derived from public-domain sources and are indicated in the Data and Maps for ArcGIS source guide as freely redistributable with attribution to the source.

The author gratefully acknowledges the publishers of the datasets that accompany this book. Special thanks are awarded to Dr. Charles R. Bacon, who created a masterful geologic map of the truly remarkable place called Crater Lake, and with whom I was privileged to work.

Data sources include the following:

- *Crater Lake National Park Map*, 2011, National Park Service. (Download) https://www.nps.gov/hfc/cfm/carto-detail.cfm?Alpha=CRLA (July 2017).

- Data and Maps for ArcGIS (airports, cities_dtl, counties_dtl, highways, parks, hydrolyn, volcanoes; subset to Oregon), 2012–2016, Esri, Redlands, California. (DVD/Download) (July 2017).

- *Geologic Map of the Edwards Aquifer Recharge Zone*, South Central Texas, 2005 (PDF), Charles D. Blome, Jason R. Faith, Diana E. Pedraze, George B. Ozuna, James C. Cole, Allan K. Clark, Ted A. Small, and Robert R. Morris, US Geological Survey Scientific Investigations Map 2873, Scale 1:200,000. (Download) http://pubs.usgs.gov/sim/2005/2873/pdf/2873_Print.pdf (June 2010).

- *Geologic Map of Mount Mazama and Crater Lake Caldera*, 2008, David W. Ramsey, Dillon R. Dutton, and Charles R. Bacon, US Geological Survey Scientific Investigations Map 2832. (Download) https://pubs.usgs.gov/sim/2832 (July 2017).

- Landsat image LT05_L1TP_045030_20110909_20160830_01_T1, 2011, EarthExplorer, US Geological Survey. (Download) http://earthexplorer.usgs.gov (November 2016).

- National Elevation Dataset 1/3rd arc-second resolution, 2016, US Geological Survey, Reston, Virginia. (Download) https://viewer.nationalmap.gov/basic (July 2017).
- National Elevation Dataset 1/9th arc-second resolution, 2016, US Geological Survey, Reston, Virginia. (Download) https://viewer.nationalmap.gov/basic (July 2017).
- *Soils Data for the Conterminous United States Derived from the Natural Resources Conservation Service (NRCS) State Soil Geographic (STATSGO) Data Base*, edition 1.1, 1995, G. E. Schwarz and R. B. Alexander, USGS Open File Report 95-449, US Geological Survey, Reston, Virginia. (Download) http://water.usgs.gov/lookup/getspatial?ussoils (July 2017).

Index

2D and 3D navigation, 44–46

3D Analyst, 5, 34

3D data, 154

3D scenes, 50–52

Active Template mode, 150

Add Clause button, 65, 79

Add Data button, 47

Add Data drop-down button, 22

Add Item function, 31

Add Join tool, 85

All Portal tab, 30

Analysis tab, 73

animation, 154

annotation, 7, 55, 136, 146–47: editing, 148–49; features, creating,
    149–51

Appearance tab, 27, 54, 58, 69–70

ArcCatalog, 1, 8, 111, 112, 113

ArcGIS Desktop, 6

ArcGIS Online, 3, 44, 153: accessing maps and data from, 14–16; account,
    3–4, 10; default locator, 35–36; sharing data using, 128–29

ArcGIS Pro: advanced capabilities, 153–54; background on, 1–2;
    capabilities of, 4–5; GUI, 1–2, 5, 12–14, 21–31; help files, 18–19;
    licensing, 3–4, 10; offline use of, 3; project creation, 11–12;
    projects, 33–42; reasons to switch to, 5–8; starting, 10–11; system
    requirements, 2–3

ArcInfo Workstation, 6, 111

ArcMap, 1, 2, 5–6, 8, 23, 43, 54, 111: data frames, 97; extensions, 154;
    importing map documents from, 19–20; Label Manager, 55, 66;
    properties in, 41; Table of Contents, 13, 25; windows in, 17

ArcScene, 4, 34–35, 44

ArcToolbox, 71

ArcView, 6

attributes: creating maps from, 60–63; selections based on, 78–80

Attributes pane, 138, 142

attribute tables, 83

background geoprocessing, 8

Band Combination button, 56, 69–70

Basemap button, 48

basemaps, 28

bookmarks, 47

broken paths, 37

Buffer tool, 72, 74–76

buttons, 22

Calculate Field tool, 95–96

cardinality, 85–87

Catalog pane, 13, 14, 16, 22, 29–31, 47, 110, 112

Catalog view, 13, 16, 17, 23, 26, 30, 112

Catalog window, 13

Center option, 44

charts, 87–88, 90–92

Charts tab, 88

class breaks, 62–63

Classes tab, 62

clip setting, 7

COGO, 4

construction tools, 133, 140–41

Contents pane, 13, 15, 16, 22, 24–27, 29, 45, 46, 51, 52, 55, 68, 84, 98–100, 131

contextual tabs, 21, 27–28, 54–56, 58, 99

Contour tool, 73

Convert Labels To Annotation tool, 147–48

coordinate systems, 6, 44, 119–21

Create Feature Class tool, 121

Create Features pane, 133, 135, 139, 140, 142, 149–51

CSV file format, 7

Curved Annotation tool, 150

custom workflows, 154

data: 3D, 154; accessing, 14–16; adding to maps, 47; downloading, 8–10; sharing, 128–29

data frames, 23, 97

data management: data models and, 111–12; from diverse sources, 116; domain creation, 115; feature classes, 120–22; fields and domains, 124–26; geodatabase schema, 112–14; metadata, 122–23; project longevity and, 116–17; shared data for work groups, 117–18; sharing data, 128–29; table schema, 126–27

data models, 111–12

data sources, 155–56: diverse, 116; links to, 36–37

Data tab, 27, 54

datums, 6

design changes, 112–14

destination table, 84

docking icon, 17

domains, 115, 124–26

Domains tab, 115

Domains view, 24, 115

DRA button, 56

drop-down buttons, 22

dynamic constraints, 134

dynamic labels, 5, 136

editing: annotation, 136, 148–49; annotation feature class, 146–47; annotation features, 149–51; basic functions, 131–32; context menus, 143; existing features, 134–35, 144–45; feature creation, 133–34; layouts, 98–100; lines, 140–42; points, 138–39; polygons, 142–43; shortcuts, 139; tables, 95–96; tools, 136–38; topological, 135

editing changes, 112, 113

Edit tab, 132, 138

Element pane, 105, 109

elevation sources, 45, 46

Environment settings, 72, 73

Environments tab, 72, 120

Esri GeoEnrichment service, 44

Excel, 6, 7

Explore tool, 21–22, 43, 44, 46–48, 50–52, 69, 139

Export Features, 120

extensions, 3, 5, 73, 154

extent, 103–4

Extent settings, 49

fault symbols, 141–42
Favorites tab, 31
feature classes, 120–22, 124, 146–47
feature layers, 14
Feature Layer tab, 28, 29, 54, 58, 115
features: annotation, 149–51; creating, 133–34, 149–51; modifying existing, 134–35, 144–45
Features group, 132
Federal Geographic Data Committee Content Standard for Digital Geospatial Metadata (FGDC-CSDGM), 111–12
fields, 124–26
Fields tab, 114
Fields view, 24, 89, 113–14, 125–26
Find tool, 43
folder connections, 19, 31, 35, 37, 39, 47, 100
folders, project, 34, 37, 39
Format labels, 62
Format pane, 104–5
Format Point Symbol mode, 58, 60
Format Polygon Symbol mode, 22
Freehand tool, 133
Full Extent button, 49–50
functions, 83

Gallery tab, 22, 58, 59
geodatabases, 35, 39
geodatabase schema, 112–14
GeoNet website, 153
geoprocessing, 7–8: Analysis buttons and tools, 73; interactive selections, 76–78; interface, 74–76; overview of, 71–73; practicing, 81–82; results output, 72; selections based on location, 80–81; selections by attributes, 78–80; tool licensing, 73–74
geoprocessing history, 35
Geoprocessing pane, 14, 71–76, 81–82, 93
geostatistical analysis, 154
global scenes, 44
Graduated Colors map, 61
Graduated Symbols map, 63

greenness index map, 70
Groups tab, 30
GUI: background on, 21–24; design of, 1–2, 5; exploring, 12–14, 24–31; panes, 22–23; ribbon and tabs, 21–22; views, 23–24
guides, 103–4

halos, 23, 67
help files, accessing, 18–19
histograms, 90–92
Histogram tab, 62, 63
History button, 73
History tab, 31
home geodatabase, 35, 39

imagery layer, 14
Imagery tab, 57, 70
ImportLog folder, 34
Index folder, 34
Infographics tool, 43–44
Inquiry group, 43–44
Insert tab, 15, 16, 40, 100, 102
interactive selections, 76–78
International Organization for Standardization ISO 19115, 112

joins, 84–87
join table, 84

label classes, 64–66
Label Class group, 23
Label Class pane, 55, 67
labeling properties, 54–55
Labeling tab, 23, 27, 54–55, 64–67
Label Manager, 55, 66
Label Placement group, 64
labels, 146–48: creating, 63–66; managing, 66–67
layer file, 34
layer metadata, 123
layer properties, 26, 37, 42, 113, 127

layers: accessing symbol settings for, 53–54; disabling editing for, 131; labeling properties, 54–55; raster, 68–70; selecting, 77–78

layout files (.pagx), 100

layouts, 20, 35: accessing and copying, 110; creating maps for, 101–2; editing procedures, 98–100; exploring, 101–10; importing map documents and templates, 100; legend, 108–10; map elements, 106–8; map frames and, 97–98, 102–6

Layout tab, 99

layout views, 24

legend, 108–10

licensing, 3–4, 10, 73–74

lines, 140–42

Link Views function, 44

List By Drawing Order tab, 83, 88

List By Editing tab, 131–32, 137

List By Labeling tab, 66

List By Selection tab, 77–78

List By Snapping tab, 132, 137

List By Source tab, 83

Living Atlas tab, 30

local resources, 29

local scenes, 44

Locate tool, 49

locators, 35–36

Make Feature Layer tool, 81

many-to-one cardinality, 85–86

map books, 153

map documents, importing, 19–20, 100

map elements, 106–8

map frames, 97–98, 102–3: formatting, 104–6; setting extent and scale, 103–4

Maplex, 4, 54, 63, 67

map properties, 52

maps, 34: 2D and 3D navigation, 44–46; accessing, 14–16; adding data to, 47; basemaps, 28; creating for layouts, 101–2; creating from attributes, 60–63; creating new, 40; exploring, 43–52; legend, 108–10; linking views, 50–52; properties of, 41–42;

symbolizing, 53–70; web, 15–16; map series, 153; map surrounds, 98

Map tab, 78

map templates, 100

Map tools, 46–50

map topology, 132, 135, 145, 153–54

map views, 17, 23

Measure tool, 43, 48

merge rules, 86

metadata, 13, 15, 42, 111–12, 122–23

mobile devices, 153

ModelBuilder, 4

Modify Features pane, 134–35, 144

My Content tab, 30

Navigate group, 22, 43

navigation, 44–46

New Map button, 40

normalized difference vegetation index (NDVI), 57, 70

north arrows, 102

objects, working with, 27–29

one-to-many cardinality, 85–87

one-to-one cardinality, 85, 87

Options dialog box, 66–67

Options menu, 23

packages, 128–29

panes, 13, 21, 22–23: arranging, 17–18

Parameters tab, 72

paths: broken, 37; in projects, 36–37

planar topology, 153–54

points, 138–39

polygons, 142–43

Portal folder, 13

portals, 36

Portal tab, 29–30

Primary symbology tab, 62–63

project files, 36

project folders, 34, 37, 39

project geodatabase, 35

projects, 10, 33–42: creating, 11–12; creating and exporting data to, 118–20; defined, 33–34; elements of, 38–40; exploring, 38–42; items stored in, 34–36; longevity of, 116–17; packaging, 128–29; paths in, 36–37; properties of, accessing, 41–42; renaming, 37–38; shared data, 117–18
Project tab, 29, 41
properties, 26, 41–42
properties changes, 112, 113
Properties tab, 22, 58, 59
Python, 4

Quick Access Toolbar, 16, 118

Raster Layer tab, 56–57, 68–70
rasters, 6, 45: symbolizing, 56–57, 68–70
Recent heading, 71
Rectangle tool, 77
renaming projects, 37–38
Rendering group, 56
Reshape tool, 134–35
ribbon, 21–22, 84

Save As command, 37–38
scale, 103–4
Scale Bar tab, 99
scale-based symbology, 61
Scale option, 44
Scales tab, 62
scenes, 34–35
scene views, 23
schema, 112–14
schema changes, 112–14, 126–27
Search icon, 18
Search window, 71
secondary tabs, 22–23
Select By Attributes button, 43
Select By Location button, 43, 80–81
Selection group, 43, 76–78, 132
Select Layer By Attribute tool, 5, 78–80
Select tool, 43, 48
shapefiles, 111

Sketch menu, 133, 134
snapping, 132, 137
source metadata, 123
source table, 84
Spatial Analyst, 5
spatial join(s), 8, 85–87
Spatial Join tool, 85, 86
standalone tables, 83
Standard Label Engine, 67
Statistics menu item, 83, 92–94
stream digitizing, 133
styles, 35, 65, 100, 153
Summary Statistics tool, 83, 93–94
Swipe tool, 69
Symbology pane, 14, 17–18, 22, 23, 29, 53–54, 58, 59, 61, 62, 63, 68, 141
symbols: fault, 141–42; layers, 53–55; modifying single, 57–60; rasters, 56–57, 68–70; scale-based, 61; settings, accessing, 53–54
symbol styles, 7
system requirements, 2–3

Tab key, 124
Table of Contents, 13, 25
tables: calculating and editing, 95–96; calculating statistics for, 92–94; cardinality, 85–87; charts and, 87–88, 90–92; general characteristics of, 83–84; joining and relating, 84–87; managing table views, 88–90
table schema, 126–27
table views, 23–24
tabs, 21–22, 27–29
target table, 84
tasks, 36
templates, 100, 133
Terrain3D, 45
time analysis, 154
toolboxes, 35
toolbox file, 34
tool licensing, 73–74
topology, 135, 153–54

Unique Values map, 60–61

Vector Field map, 68
Vertex menu, 134
Vertices tool, 144–45
views, 13, 21, 23–24: linking, 50–52; table, managing, 88–90
View tab, 44, 50

web maps, 15–16
windows, 13: arranging, 17–18
work groups, shared data for, 117–18
WorldElevation3D/Terrain3D, 45
World Geocoding Service, 35–36

XML format file, 112

zooming, 47